MECHANISMS OF GENETIC RECOMBINATION

STUDIES IN SOVIET SCIENCE

LIFE SCIENCES

1973

Motile Muscle and Cell Models
N. I. Arronet

Pathological Effects of Radio Waves
M. S. Tolgskaya and Z. V. Gordon

Central Regulation of the Pituitary-Adrenal Complex
E. V. Naumenko

1974

Sulfhydryl and Disulfide Groups of Proteins
Yu. M. Torchinskii

Organ Regeneration: A Study of Developmental Biology in Mammals
L. D. Liozner

Mechanisms of Genetic Recombination
V. V. Kushev

A Continuation Order Plan is available for this series. A continuation order will bring delivery of each new volume immediately upon publication. Volumes are billed only upon actual shipment. For further information please contact the publisher.

STUDIES IN SOVIET SCIENCE

MECHANISMS OF GENETIC RECOMBINATION

V. V. Kushev
A. F. Ioffe Physicotechnical Institute
Academy of Sciences of the USSR
Leningrad, USSR

Translated from Russian by
Basil Haigh

CONSULTANTS BUREAU • NEW YORK AND LONDON

Library of Congress Cataloging in Publication Data

Kushev, Vladislav Valer'ianovich.
  Mechanisms of genetic recombination.

  (Studies in Soviet science)
  Translation of Mekhanizmy geneticheskoĭ rekombinatsii.
  Bibliography: p.
  1. Genetic recombination. I. Title. II. Series.
  QH443.K8813   575.1   73-83898
  ISBN 0-306-10891-7

Vladislav Valer'yanovich Kushev was born in 1939 in Leningrad. In 1963 he was graduated from the Department of Genetics, Leningrad University. He has specialized in the study of the mechanism of genetic recombination in transformation of bacteria and is the author of more than ten publications on this subject. He is now engaged in research at the Laboratory of Biopolymers at the Leningrad Institute of Nuclear Physics.

The original Russian text, published by Nauka Press in Leningrad in 1971, has been corrected by the author for the present edition. This translation is published under an agreement with the Copyright Agency of the USSR (VAAP).

МЕХАНИЗМЫ ГЕНЕТИЧЕСКОЙ РЕКОМБИНАЦИИ
В. В. Кушев
MEKHANIZMY GENETICHESKOI REKOMBINATSII
V. V. Kushev

© 1974 Consultants Bureau, New York
A Division of Plenum Publishing Corporation
227 West 17th Street, New York, N.Y. 10011

United Kingdom edition published by Consultants Bureau, London
A Division of Plenum Publishing Company, Ltd.
4a Lower John Street, London W1R 3PD, England

All rights reserved

No part of this book may be reproduced, stored in a retrieval system, or transmitted, in any form or by any means, electronic, mechanical, photocopying, microfilming, recording, or otherwise, without written permission from the Publisher

Printed in the United States of America

> "Recombination is a basic biological process; to understand its molecular basis is clearly a tough and exciting challenge."*
>
> G. Pontecorvo

Preface

"Genetic recombination" is the name given to the redistribution of information inherited from the parents in the progeny. Although this definition applies to the recombination of both linked and nonlinked genes, the term "recombination " is usually applied in a narrower sense, to mean only recombination of linked markers (between different genes and within the same gene). It is in this sense that the term is used both in the title of the book and in the text (except Section 1.1).

Recombination of nonlinked genes is based on a mechanism of free combination of chromosomes in meiosis. Recombination of linked genes usually takes place by crossing-over of chromosomes. Recombinations within the same gene are in most cases (if not always) the result of conversion of one of the alleles into the other, a process known as gene conversion. In crossing-over there is a reciprocal exchange of information between homologous chromosomes, while in conversion this exchange is predominantly nonreciprocal in character.

Genetic recombination is a fundamental property of all living systems starting from the RNA-containing viruses (Kawai and Hanafusa, 1972) and ending with higher plants and animals. As the source of variation through combination, genetic recombination is one of the principal factors in evolution. At the same time, however, results showing that genetic recombination is one of the main sources of variation through mutation have also been described. Recombination mechanisms, according to some investigators, may also lie at the basis of variation within the organism and to take part in such processes as the formation of antibodies, of communicative proteins, and so on. The link discovered recently between mechanisms of recombination and cell repair has converted the study of the mechanism of genetic recombination from a purely internal problem of genetics into a

---

Genetics Today, Vol. 2, New York and London, p. 89.

general problem of molecular biology. Enzyme systems responsible for recombination have been found to take part in such fundamental processes as the correction of mistakes in the genetic text. It is not surprising, therefore, that the serious study of the molecular mechanism of recombination should have begun only a few years ago. Not until more elementary matters had been explained was it possible to make an approach to this complex problem. To begin with, the structure of the elements interacting during recombination (DNA molecules and chromosomes) had to be established, and the method of their replication had then to be explained; the biological properties of many new parasexual systems (bacteriophages; conjugation, transformation, transduction, sexduction in bacteria) had to be carefully studied; if not the nature of chromosome-pairing processes, at least the moment of conjugation of the chromosomes in meiosis had to be determined; the structure of the recombinant molecules studied; and, finally, the character of the recombination process discovered. All this had to be done, of course, on a basis of the achievements of classical genetics, which had yielded strict and final proof of the chiasmatype theory.

However, despite evident and rapid progress, we are still only halfway to our goal. During the crisis of the naively optimistic views on the nature of recombination (the mechanical crossing-over hypothesis), which began with "splitting" of the gene, new facts were discovered which could not be explained fully at that time by contemporary physicochemical and biochemical concepts of crossing-over. Facts of this type included high negative interference, gene conversion, the polarity and allele specificity of recombination within the gene, the nonadditive and even linear character of the ultrashort distances, and so on. Proof that recombination within the gene is intimately connected with crossing-over indicates that it takes place at the sites of breaking and rejoining of the chromosomes. In this connection the facts obtained by the study of recombination within the gene, especially by the use of tetrad analysis, provide extremely valuable material for a construction of a general theory of crossing-over.

Many thousands of investigations have been undertaken in order to study the mechanism of genetic recombination, starting with the work of Morgan and continuing until the present time. The flow of publications has now increased so rapidly and the fields of research have widened so considerably that there is now an absolute need for a survey of the problem *in toto*. An attempt is made in this book to generalize the results obtained in genetic systems of different complexity from phages to mammals. This approach may appear unduly rash, but the universality of the fundamental metabolic processes revealed by molecular biology inspires confidence that the fundamental mechanisms of genetic recombination are also universal for the whole living kingdom. Specific differences undoubtedly exist, but it

PREFACE

will be more useful to find the general principles first and then to go on to explain the exceptions and to define the limitations.

At the same time the book cannot claim to be an exhaustive survey of the literature which has accumulated since the publication of the last monograph on this problem (Stern, 1933). Moreover, many of the facts and hypotheses which, in the present writer's opinion, lie outside the mainstream of development of ideas on the mechanism of genetic recombination have simply been omitted. No attempt likewise is made to analyze special cases of genetic recombination such as lysogenization the formation of specialized transducing phage particles and F' factors; transposition of the dissociating element (Ds) in corn, recombination of the factors determining the type of conjugation in *Schizophyllum commune,* etc.

The principle used in the planning of this book, progressing from facts to hypotheses on the historical plane, must certainly make it more difficult to read because it is much easier to examine facts from the standpoint of a preexisting scheme. However, since no general theory of crossing-over has yet been developed, the book must inevitably be of this character. Modern theories of crossing-over are described only in the last chapter, Chapter 5. This chapter is entirely theoretical and speculative in character. The fundamental facts on which the corresponding speculations are developed are described in the first four chapters.

In Chapter 1 the birth and development of recombination analysis are briefly described with respect to recombination between genes. The chiasmatype theory of Janssens and Morgan is examined in detail, together with alternative hypotheses for the mechanism of crossing-over; breakage and reunion and copy-choice both in the classical form and in its modern modifications.

Facts leading to a crisis in the classical views of the mechanism of recombination and also a number of hypotheses put forward in an attempt to explain these facts within the framework of modified classical ideas are considered in Chapters 2 and 3.

The structure of chromosomes, the method of their replication, and identification of the stages of DNA synthesis, of crossing-over, and of conversion in meiosis are discussed in Chapter 4. Modern physicochemical, genetic, and biochemical results supporting the mechanism of breaking and rejoining and of the participation of enzyme systems in recombination are given. The possible mechanisms of induced crossing-over and of structural changes in chromosomes are also discussed.

Facts obtained by the study of recombination within the gene are interpreted in Chapter 5, and the problem of molecular heterozygosis is analyzed. The two principal groups of models of recombination are discussed. The first group includes models with statistical breaks and exonucleotic repair, the second group models with fixed points of breaks and endonucleotic repair (correction). In my view the process of recombination can be adequately described by adoption of the principle of directed correction.

Many readers will not find here a detailed analysis of the systems with which they are working. However, this book would never have been written if I had not limited my task to the examination of only the most general problems. It likewise would never have seen the light of day without the help, support, and good advice of many people to whom I am deeply grateful: I. A. Zakharov, M. I. Mosevitskii, R. A. Kreneva, V. L. Kalinin, Z. V. Kusheva, I. E. Vorobtsova, and, in particular, S. E. Bresler.

I am indebted to Plenum Press for allowing me to update the text with material obtained in the last two years, which has added about 300 new titles to the bibliography. I warmly acknowledge the interest shown by R. Holliday and J. Scaife in the progress of the English edition of this book.

V. V. K.

Leningrad
June, 1973

Contents

Chapter 1. Recombination of Genes ... 1
  1.1 Mendelism ... 1
  1.2 Sutton's Prophecy ... 2
  1.3 The First Hypothesis of Recombination of Linked Genes ... 5
  1.4 The Chiasmatype Theory of Janssens ... 5
  1.5 Morgan's Experiments on *Drosophila* ... 7
  1.6 The Genetic Map ... 7
  1.7 Interference ... 11
  1.8 The Physical Scale of Genetic (Linkage) Maps ... 13
  1.9 Chromatid Crossing-over ... 14
  1.10 The Problem of Chromatid Interference ... 17
  1.11 Chiasmata and Crossing-over ... 20
  1.12 Cytological Evidence for Crossing-over ... 21
  1.13 Darlington's Hypothesis ... 22
  1.14 Belling's Hypothesis ... 23
  1.15 Sister-Strand Exchanges ... 24
  1.16 Mitotic Crossing-over ... 27
  1.17 Merozygote Systems ... 27
  1.18 Recombination in Bacteriophages ... 31
  1.19 The Copy-Choice Hypothesis ... 33
  1.20 Conclusion ... 35

Chapter 2. Intragenic Recombination. Random-Sample Analysis ... 37
  2.1 The Concept of the Gene ... 37
  2.2 Step Allelism ... 39
  2.3 Pseudoallelism ... 39
  2.4 Intragenic Recombination ... 41
  2.5 The Structure of the Gene ... 42
  2.6 The Recon ... 46

2.7	Additivity within the Gene	50
2.8	The Flank Marker Method	53
2.9	High Negative Interference	56
2.10	The Effective Pairing Hypothesis	58
2.11	Recombinational Discontinuity of the Chromosomes	61
2.12	Allele Specificity of Recombination	63
2.13	Conclusion	69

Chapter 3. Gene Conversion. Tetrad Analysis of Intragenic Recombination ... 71

3.1	History of the Problem	71
3.2	The Conversion Hypothesis of High Negative Interference	74
3.3.	Mitchell's Discovery	75
3.4	Crossing-over or Conversion?	76
3.5	The Switch Hypothesis	81
3.6	The Polaron Hypothesis	82
3.7	The Relationship between Conversion and Crossing-over	88
3.8	One or Two Mechanisms of Genetic Recombination?	90
3.9	Origin of Tetrads Containing Reciprocal Products of Recombination between Alleles	96
3.10	Conclusion	101

Chapter 4. Experimental Study of the Mechanism of Recombination .. 103

4.1	Introduction: The Problem Stated	103
4.2	The Structure of Chromosomes	104
4.3	The Character of Replication of DNA and Chromosomes	110
4.4	Identification of the Stage of DNA Synthesis in Meiosis	111
4.5	At What Stages of Meiosis Do Conversion and Crossing-over Take Place?	113
4.6	Synapsis of Chromosomes	117
4.7	Structural Aspects of Stadler's Modified Polaron Hypothesis	121
4.8	Breakage and Reunion	123
4.9	The Biochemistry of Crossing-over	132
4.10	The Mechanism of Induced Recombination	144
4.11	Conclusion	151

Chapter 5. Modern Theories of Genetic Recombination ... 153

5.1	Introduction	153
5.2	Random Breakage Models	154
5.3	Taylor's Model	157
5.4	Holliday's Model	159
	Polarity	159
	Correction of Molecular Heterozygosity	160

	High Negative Interference (Interpretation of Distribution of Flank Markers)	164
	Nature of Reciprocal Intragenic Recombination	167
5.5	The Whitehouse–Hastings Model	170
	Polarity	170
	Origin of the t Tetrads	172
	Interpretation of Flank Classes	174
5.6	A Theory of Genetic Recombination Based on the Principle of Directed Correction	176
	The Principle of Directed Correction	176
	The Model	177
	The Origin of t Tetrads	178
	Evidence for the Directed Correction Hypothesis	179
	Distribution of Flank Markers	181
	Polarity	185
	Linear Measurements	190
	Expansion of the Map	192
	Allele Specificity of Recombination	193
	The Genetic Control of Intragenic Recombination	199
	Heterozygosity in Bacteriophages	200
	Mechanisms of Recombination in Merozygous Systems	210
	High Negative Interference in Three-Point Crosses	222
	Correction and Repair	223
5.7	Conclusion	226

Bibliography 229

CHAPTER 1

Recombination of Genes

1.1. Mendelism

The analytical approach to the problem of heredity, which was first adopted by Mendel (1866), led him to consider that there are certain discrete factors (subsequently called genes) which are transmitted from generation to generation through the sex cells and which determine the development of the inherited characters of the organism. This hypothesis was suggested by Mendel in order to interpret the surprising rules governing inheritance of characters which he had discovered in his experiments on *Pisum sativum*. When two plants differing from each other in a pair of characters which are inherited constantly in the case of self-pollination [for example, seed color: one plant with yellow seeds (A), the other with green seeds (a)], only yellow seeds are formed in the first generation (F_1). However, although one of the characters has apparently disappeared in F_1, the genetic factor determining it has not disappeared and is not "dissolved" in the hybrid, but its presence is revealed in crossings of the type $F_1 \times F_1$ or $F_1 \times P(a)$, where $P(a)$ denotes the parental form with green seeds. Crossings of this last type are called analytical. Segregation of the pattern $3A:1a$ is found in the progeny from the first cross and $1A:1a$ in the progeny from the second.

If the parental (P_1 and P_2) plants differ in two pairs of alternative characters [for example, in the color of the seeds, as in the previous crossing, and in their shape (B, smooth peas; b, wrinkled peas), so that P_1 has the phenotype AB and P_2 the phenotype ab], forms with the parental phenotypes (AB and ab) as well as new phenotypes (Ab and aB), combining characters of both parents, will be formed in analytical crossing in equal numbers ($1:1:1:1$). The last two are called recombinants (R_1 and R_2). With respect to each character separately, a $1:1$ segregation is observed in analytical crossing and, consequently, the characters are inherited independently of each other, they are combined at random in the progeny, and

the resulting segregation is obtained by superposition of the two ratios
$1A : 1a$ and $1B : 1b$.

The ratio between the number of recombinants $(R_1 + R_2)$ and the total number of progeny is a measure of the recombination frequency (rf) of the characters; rf is expressed either as a fraction or as a percentage:

$$rf = \frac{R_1 + R_2}{P_1 + P_2 + R_1 + R_2} (100).$$

According to Sturtevant (1965) this equation was suggested by H. Muller. In the case of independent inheritance of the characters $rf = 0.5$ (50%).

Mendel investigated seven pairs of alternative characters in peas and showed that they are all inherited independently of each other. In the first few years after the rediscovery of Mendel's laws, the principle of independent inheritance of characters seemed to be universal. However, later genetic research together with the quest for the cytological mechanisms lying at the basis of the principles of inheritance have compelled revision of this idea.

1.2. Sutton's Prophecy

At the time of the rediscovery of Mendel's laws, the nucleus-chromosome hypothesis of inheritance (Weismann, 1885) had already been formulated. In 1902 this hypothesis received its first confirmation by the cyto-embryological investigations of Boveri (1902), who demonstrated the individuality and the differential role of the chromosomes in the inheritance and expression of characters. In this connection cytologists were compelled to pay the closest attention to the study of concrete processes taking place in the cells and their chromosomal apparatus during both mitosis and meiosis (Fig. 1).*

---

* Meiosis consists of two successive divisions of a diploid cell with the formation of four haploid cells (gametes). The first division is a reduction division, for as a result of it the number of chromosomes is halved. The second division is usually called an equation division. However, if crossing-over takes place, the sister chromatids, which separate only during division of the centromere in the anaphase of the second division, are not identical, and for this reason segregation is not complete until the second division (1.4).

The behavior of the chromosomes in prophase I, which is divided into several stages, is specific for meiosis. In the leptotene stage the thinnest chromosome threads become visible under the light microscope. Each of them as a rule appears to consist of only one chromatid, despite the fact that at this time the DNA has already replicated (4.4). The duplex nature of the chromosomes at this stage can be recognized only in the electron microscope. In the zygotene stage the chromosomes are thicker and shorter, and homologous chromosomes pair with each other (synapsis). Synapsis or conjugation usually begins at one end of a pair of homologues and spreads like a zipper to the other end. After conjugation, this pair of homologues is called a bivalent. In the pachytene stage the con-

Sutton (1902), who studied meiosis in the cricket *Brachystola magna*, directed attention to the surprising similarity in behavior of the chromosomes during meiosis and gene recombination, from which he drew conclusions regarding their chromosomal localization. These ideas were described in more detail in a later paper (Sutton, 1903). Sutton also pointed out such distinctive features of meiosis as the pairing of homologous chromosomes, one of which is maternal and the other paternal in origin; the union of the homologous chromosomes during conjugation; the reduction character of the first and equation character of the second division; the preservation of morphological individuality of the chromosomes in the division cycle; and the independence of behavior of one pair of homologous chromosomes of the behavior of other pairs, a fact which evidently lies at the basis of independent inheritance of different pairs of characters. The subsequent development of the idea of the link between chromosomes and genes (in Sutton's terminology, allelomorphs) led him to the prophetic conclusion that the law of independent inheritance has limitations: if the basis of one allelomorph is not the whole chromosome but only part of it, and "if chromosomes constantly retain their individuality, all allelomorphs present in one chromosome must be inherited together" (Sutton, 1903; p. 235). In Sutton's opinion, therefore, the action of the law of independent inheritance extends only to genes located in different chromosomes; consequently, the number of freely (independently) recombining genes is limited to the number of chromosomes in each individual organism.

Sutton predicted that the problem of limited application of this law of Mendel must arise wherever the number of genes studied in a single organism exceeds the haploid number of chromosomes which it possesses. However, Sutton's prediction apparently did not materialize; only seven pairs of chromosomes were known in the pea, but nevertheless besides the seven pairs of independently recombining characters studied by Mendel, a further four were discovered; this fact was quickly used as an argument against the chromosome hypothesis of inheritance.

---

jugated homologues become thicker still and the "equatorial space" becomes visible — each chromosome can be seen to consist of two sister chromatids. Recoiling of the chromosomes takes place. This has the longest duration of all the stages of prophase.

In the diplotene stage homologous chromosomes begin to be repelled from each other and the chiasmas become visible. In diakinesis the chromosomes become much shorter and thicker. The number of chiasmas is reduced through their movement toward the ends of the bivalent (terminalization). In stage I of metaphase the homologues are arranged in the equatorial plane of the division spindle, and finally, in stage I of anaphase they separate toward the poles. Either homologue moves with equal probability and at random toward either of the two poles. In addition, separation of one pair of chromosomes takes place independently of separation of the chromosomes of another pair. Separation of sister chromatids of each chromosome takes place in the second division during "splitting" of the centromere (see the survey by Rhoades, 1961).

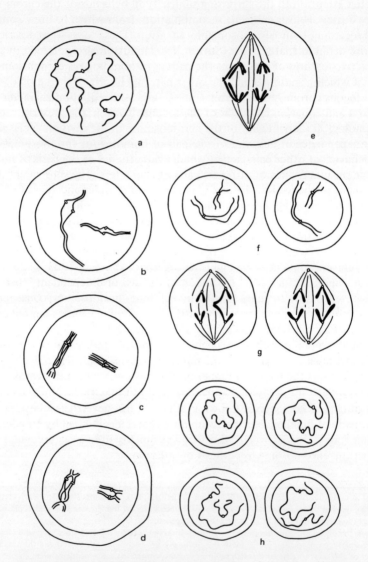

Fig. 1. Scheme of meiosis. First meiotic division: a) leptotene, b) zigotene, and c) diplotene stages, d) diakinesis, e) anaphase. Second meiotic division: f) prophase, g) anaphase, h) gametes.

1.3. The First Hypothesis of Recombination of Linked Genes

To save the chromosome theory of inheritance, Hugo de Vries (1903) postulated that homologous chromosomes can exchange small segments during their conjugation in meiosis: the alleles of one gene, lying opposite one another in homologous chromosomes, change places; the probability of such an exchange is arbitrary. It thus follows that linking (i.e., associated inheritance) of genes in one chromosome, postulated by Sutton, was found to take place in principle by genetic methods. Nevertheless, Bateson et al. (1905) found linked inheritance of characters in the sweet pea (*Lathyrus odoratus*). However, they did not interpret this fact from the standpoint of Sutton's chromosome hypothesis for the linkage found was incomplete (recombinants still appeared although only very rarely).

The fact that linkage of genes could be incomplete required new ideas for its adequate interpretation.

1.4. The Chiasmatype Theory of Janssens

At the end of last century Rückert (1892) found that homologous chromosomes joined together at various sites along their length in one of the stages of meiosis to form χ-shaped figures (chiasmata). This observation led him to postulate a possible exchange of material between chromosomes. Rückert's hypothesis was unpopular for a long time because the doctrine of the constancy of chromosomal structure prevailed among cytologists. It was revived 17 years later by Janssens (1909). On the basis of his observations on spermatogenesis in the salamander (*Batrachoseps attenuatus*) he also postulated that at sites of contact between homologous chromosomes and of chiasma formation.

The classical point of view on chiasma formation was that homologous chromatids simply twist around each other and an exchange of partners takes place at the chiasma itself (the heterotype–homotype theory) (Fig. 2A). Janssens put forward the chiasmatype theory (Fig. 2B), the essence of which is as follows. During conjugation homologous chromosomes twist around each other like the strands of a rope. They then conjugate and shorten. In the diplotene stage each chromosome splits into two and the so-called equatorial space is formed. Since the space lies in one plane, i.e., it does not correspond to the terms of the fusing chromosomes, at the sites of contact between the homologues it cuts through the chromosome and leads to breakage and rejoining of two of the four chromatids. As a result one chromatid of each chromosome is composed of parts of the two original chromosomes. On separation of the homologues these sites of breakages and rejoinings become visible as chiasmata. Janssens based his theory not only on experimental facts but also on theoretical arguments: the formation

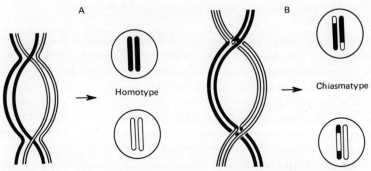

Fig. 2. Mechanism of chiasma formation: A) in accordance with the classical heterotype–homotype theory the chromosomes are coiled around each other without exchanging material. In the first division of meiosis homotypes (cells containing chromosomes of only one of the parental types) are formed from a cell combining chromosomes of the parents (heterotype). B) According to Janssens' theory, during chiasma formation two of the four chromatids exchange material; after the first meiotic division the sister chromatids are thus not identical each other (chiasmatypes) and segregation is completed only in the second division.

of four spores (or gametes), the presence of two divisions, conjugation of the chromosomes, and the formation of chiasmata — the whole of this complex mechanism of meiosis would be inconceivable if reduction occurred in the first division. In addition, complete linkage of all genes in the chromosome is of more than doubtful value from the standpoint of evolution. The necessity for some mechanism by which genes located in the same chromosome can recombine had already become evident to many geneticists. According to Janssens, however, chiasmata are formed by breakage and reunion of homologous chromosomes, resulting in recombination of genes located so far away from each other that a chiasma could form between them.

Similar views (although in a less detailed and less substantiated form) were expressed by Lock (1906): if linked characters studied by Bateson are determined by genes lying in the same chromosome, their recombination must be the result of a limited exchange of material between the homologues during conjugation. By the time that Morgan and his collaborators had begun their historic experiments on the fruit fly (*Drosophila melanogaster*), the fundamental ideas and even the experimental facts were already in existence. First, there was Sutton's idea of gene linkage; second, there was the idea of recombination of linked genes put forward by de Vries, Lock, and Janssens; and third, there was the fact of incomplete linkage of characters in the sweet pea. A sufficiently simple system was required by which the existing hypotheses could be quickly and effectively tested and new facts discovered. It was Morgan who found such a system.

1.5. Morgan's Experiments on *Drosophila*

Drosophila has proved to be an exceptionally convenient object for genetic research both for its fertility and rapidity of development and also because it has only four pairs of chromosomes. The last fact has proved particularly beneficial because it enabled several sex-linked characters to be quickly discovered and the localization of the genes determining them in the X-chromosome to be proved. In the first stages of his investigations, Morgan (1911a) found that two genes – white (determining eye color) and miniature (wing formation) – localized in the X-chromosome, are inherited almost independently ($rf \simeq 40\%$). Morgan interpreted this fact on the basis of de Vries' hypothesis. However, as other genes began to be investigated (Morgan, 1911b) it was found that, besides the cases of completely independent inheritance of genes ($rf = 0.5$) located in one chromosome, other cases of incomplete linkage exist, similar to those observed by Bateson et al. (1905) ($rf < 0.5$). Soon after this Morgan was forced to reject de Vries' hypothesis and he explained his observations on the basis of the chiasmatype hypothesis of Janssens.

In terms of the chiasmatype hypothesis variations in the strength of linkage of different pairs of genes localized in the same chromosome – from almost complete linkage ($rf \leqslant 0.5$) to independent inheritance – were explained by Morgan as follows. If each gene occupies a strictly definite locus of the chromosome, if the genes are arranged in linear order along the chromosome, and if different alleles of one gene are localized at strictly identical loci of homologous chromosomes, "in consequence" (of chiasma formation) "the original materials will, for short distances, be more likely to fall on the same side as the last, as on the opposite side. In consequence, we find coupling in certain characters, and little or no evidence at all of coupling in the other characters; the difference depending on the linear distance apart of the chromosomal materials that represent the factors" (Morgan, 1911c). Morgan and Cattell (1912) introduced the term "crossing-over" to describe recombination of linked genes. Crossing-over is a mutual (reciprocal) exchange of identical segments of homologous chromosomes leading to recombination of the linked genes.

1.6. The Genetic Map

Credit for the discovery of the fundamental law of genetic recombination belongs to Sturtevant. In his memoirs he writes: "In 1909 Castle published diagrams showing the relations between genes determining the color of a rabbit .... They were received as an attempt to discover spatial relations in the nucleus. At the end of 1911, in a conversation with Morgan about this attempt .... I suddenly understood that variations in the strength of linkage, already ascribed by Morgan to differences in the spatial arrangement of the genes, make it possible to determine the sequence in the linear dimension of the chromosome. That night I drew the first map" (Sturtevant, 1965, p. 305).

Sturtevant argued as follows. If the arrangement of the genes in the chromosome is linear and the recombination frequencies (rf) depend on the physical distance between them, the rf values must be additive. Consequently, the genes must be arranged like dots in a straight line at distances apart proportional to rf, i.e., they must form a genetic map. For example, rf between the yellow (yellow body color) and white genes is 1.3%, between the white and miniature genes 32.6%, and between the yellow and miniature genes rf = 33.8% (Morgan, 1911b). This last figure is approximately equal to the sum of the other two, from which it follows that the white gene lies between the yellow and miniature genes on the chromosome (Sturtevant, 1913). In the general case the distance between two genes is equal to the sum of or the difference between the distances between them and any third gene located in the same linkage group (the law of additivity): $rf_{ac} = rf_{ab} \pm rf_{bc}$.

Genetic maps can be drawn because of two fundamental properties of the organization of the genetic apparatus: its linear structure and the fact that recombination is functionally dependent on physical distance. The law of additivity which expresses this dependence (however simple it seems in retrospect) is one of the most important generalizations (after Mendel's laws) at the basis of modern genetics. The law of additivity shows that rf is a linear function of distance: $rf = \alpha l$, where l is the physical distance and α a coefficient of proportionality. Consequently, distances between genes on the chromosome can be expressed in recombination frequencies. These will serve as genetic distances (d). Sturtevant suggested that a segment of chromosome within which there is on the average one crossover to every 100 gametes be taken as the unit of measurement of genetic distance (d). He did not distinguish between the frequency of appearance of recombi-

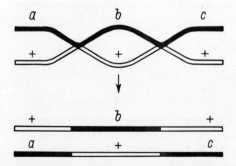

Fig. 3. Diagram of double crossing-over: a, b, c) genetic markers: +) their alleles on the homologous chromosome. If the central marker b is not taken into account, recombination between a and c will not be found. The frequency of appearance of the recombinants +b+ and a+c is a measure of the probability of double crossing-over (see 1.7).

nants in the analysis of a cross (rf) and the frequency of crossing-over, on the assumption that if recombinants with respect to a particular linked character appear in that cross with a frequency p, this means that in the section between the genes determining these characters crossing-over must have occurred in p of the 100 gametes. However, this is true only as a first approximation. Sturtevant showed that the law of additivity (in the form given above) is satisfied exactly only if neither component (rf_{ab} or rf_{bc}) is more than about 10%. On longer segments of chromosomes the experimentally determined value of rf_{ac} is always less than that expected theoretically ($rf_{ab} + rf_{bc}$). A simple logical argument shows that this must be so.

On the basis of the law of additivity, rf_{ay} can be determined in two ways: either by measuring it directly in the corresponding cross ($a \times y$) or by adding together the rf values determined at short intervals between this particular pair of genes a and y (in the presence of intermediate markers $b, c, d, ..., x$). Clearly by consecutive summation of even small distances in this manner we can obtain values for the total length of the chromosome (if a and y are located at its ends) in hundreds and thousands of percent (1.8, Table 1). Meanwhile the recombination frequency between a and y cannot exceed 50% (independent inheritance of characters). This example shows, first, that the units of measurement of distance can be regarded only conventionally as percentages and it would be better to give them another name and, second, that a mechanism exists for independent recombination of genes located in the same chromosome, but a considerable distance apart, to take place.

Deviations from additivity over long distances were ascribed by Sturtevant to the existence of double crossovers. As the diagram (Fig. 3) shows, the second crossover between a and c eliminates the results of the first: recombinants relative to a and c do not appear, although recombination took place, as can be detected by the appearance of double recombinants if an intermediate marker b is present.

Since exchanges take place randomly and at any point of the chromosome, recombinants with respect to a and c appear only when crossing-over took place in the interval 1 ($a-b$) or interval 2 ($b-c$), but not in both simultaneously. This can be expressed mathematically as follows:

$$rf_{ac} = rf_{1+2} = rf_1(1-rf_2) + (1-rf_1)rf_2 = rf_1 + rf_2 - 2rf_1 rf_2.$$

This is known as Trow's equation (Trow, 1913).

In the general case an even number of exchanges between markers does not lead to the appearance of recombinants with respect to these markers, while an odd number does. If the interval is long enough, the number of even exchanges is equal to the number of odd exchanges (recombination equilibrium) and $rf = 50\%$. In the study of long segments of the chromosome the probability of recombination is therefore no longer a measure of genetic distance. Consequently, if the distance is defined as the mean number of exchanges (d) in a certain segment of chromosome, the problem is to find d from measured values of rf, i.e., to find the mapping function. If the function is correctly determined, it must automatically give the additive values of the distances (d). The ideal mapping function can be obtained on the basis of the following arguments.

Let us assume that the points of exchange are distributed at random along the chromosome and that their mean density is constant throughout its length. Let the distance between the markers with respect to which recombinants appear with frequency rf be d. The probability that in the given distance d there are precisely n exchanges can be expressed by Poisson's formula:

$$p_n(d) = \frac{d^n e^{-d}}{n!}, \quad n = 0, 1, 2, \ldots$$

Since recombinants appear only when $n = 1, 3, 5, \ldots$, it follows that

$$rf(d) = \sum_{n=0}^{\infty} p_{2n+1}(d) = e^{-d} \sum_{n=0}^{\infty} \frac{d^{2n+1}}{(2n+1)!} = \frac{1}{2}(1 - e^{-2d}).$$

This equation was suggested by Haldane (1919). The value of d is the genetic distance between two loci and it has the required additive properties. This function can also be obtained purely mathematically from Trow's equation. Trow's equation can be rewritten in the form

$$1 - 2rf_{1+2} = (1 - 2rf_1)(1 - 2rf_2)$$

If the distance $a-c$ is broken up into n consecutive segments, Trow's equation can be easily generalized:

$$1 - 2rf_{1+2+\ldots+n} = \prod_{i=1}^{n}(1 - 2rf_i).$$

For any number of segments the term $1 - 2rf$ is a multiplicative function. Consequently the additive function can be defined as follows:

$$d(rf) = k \ln(1 - 2rf),$$

where k is an arbitrary constant and d is the frequency of exchanges between two genes between which the recombination frequency is rf. Taking k as $-\frac{1}{2}$, so that $d = rf$ at small values of the latter, we obtain Haldane's equation:

$$d(rf) = -\frac{1}{2} \ln(1 - 2rf)$$

or

$$rf(d) = \frac{1}{2}(1 - e^{-2d}).$$

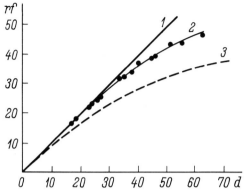

Fig. 4. Mapping functions. Abscissa, d (in %); ordinate, rf (in %). 1) Curve describing Sturtevant's function ($rf = d$); 2) empirical curve (Mather, 1938) drawn through points each of whose ordinates is the recombination frequency for a given pair of mutants, while the abscissa is the sum of the recombination frequencies for small consecutive intervals (d) between that pair of markers; 3) curves describing Haldane's function.

Haldane's function shows that as the distance increases the value of rf approaches 0.5 asymptotically. The values of rf and d for practical purposes coincide only on the segment of the curve below $rf = 0.1$ (Fig. 4).

1.7. Interference

Haldane set out from the assumption that exchanges take place independently of each other and are distributed along the chromosome in accordance with Poisson's function. However, Muller (1916), who analyzed the experimental results obtained by Sturtevant and Morgan, found that simultaneous crossovers are not independent. Double crossovers take place at a frequency below that expected theoretically (i.e., the product of the probabilities of single exchanges). For example, if $rf_{y\text{-}w} = 1.3\%$ and $rf_{w\text{-}mi} = 32.6\%$, the expected frequency of double exchanges is 0.43%. Yet the observed frequency is only 0.045%. This must mean that a crossover which has taken place reduces the likelihood of a second crossover close to it. Muller called this phenomenon interference[*] and suggested that its intensity

---

[*]Since interference of this type is based on interference between chiasmata (1.11), the name "chiasma interference" has been given to it (Mather, 1933a). However, since chiasmata, in the classical meaning of the term, have not been found in microorganisms, the use of the term "chiasma" in this connection is open to objection. A better term is evidently chromosome interference. At the level of phenomenological description, simply the term "interference" is used. If $C < 1$, interference is called positive, while if $C > 1$, it is negative.

be measured as the coincidence C:

$$C = \frac{\text{experimentally determined frequency of double crossover}}{\text{theoretically expected frequency of crossover}}.$$

In the example given $C = 0.106$. C can also be determined by a modified Trow's equation:

$$rf_{1+2} = rf_1 + rf_2 - 2Crf_1rf_2,$$

whence

$$C = \frac{rf_1 + rf_2 - rf_{1+2}}{2rf_1rf_2}.$$

This equation is more suitable because it does not require the setting up of triple-factor crosses. Muller showed that C depends on distance: the greater the distance the higher the value of C. If the distance is great enough (35%) interference disappears ($C = 1$).

The curve describing Haldane's function in Fig. 4 clearly deviates sharply from linearity from the very beginning (from $rf = 0.1$). The presence of interference compensates for the effect of multiple crossing-over, so that the experimental curve remains linear up to $rf = 0.25$. The intensity of interference varies considerably in different species and even in different parts of the same chromosome. For example, near the centromere of all chromosomes of *Drosophila* $C = 0$; across the centromere $C = 1$ (Stevens, 1936). In *Aspergillus nidulans* $C = 1$ for all regions of all chromosomes, i.e., interference is totally absent or even very slightly negative (Strickland, 1958a, 1961). The same situation is found also in *Ascobolus immersus* (Paszewski et al., 1966). Conversely, in mice interference is much stronger than in *Drosophila* or corn (Parsons, 1958). The intensity of interference is evidently under genetic control. Mather (1936a, 1937) showed by cytogenetic investigations that chiasmata are formed in a regular order beginning from the centromere. The first chiasma formed differs from the rest in that the mean interval between it and the centromere (called the differential distance) correlates with the length of the chromosome arm, whereas subsequent chiasmata form intervals whose mean length (the interference distance) is constant for all chromosomes of the same organism.

The form of the particular mapping function for the conversion of rf into d thus depends on the laws controlling interference in the organism concerned. If these laws are known (*Drosophila*) the function can be plotted (Kosambi, 1944; Owen, 1950). The criterion of the adequacy of these functions is the additivity of d: if the function is properly determined the values of d must be additive for all values of rf. The problem of strictly accurate determination of the mapping function is in many cases one of considerable difficulty (Mather, 1951; Stahl et al., 1964), and for this reason in

TABLE 1. Comparison of Genetic and Physical Dimensions of Chromosomes in Different Organisms

Organism	Length of all linkage groups, in map units	DNA content in genome, in base pairs	Number of base pairs per map unit	Author
Bacteriophage T4	2000	$2 \cdot 10^5$	$1.0 \cdot 10^2$	Doermann and Parma, 1967
E. coli	2000	1.0×10^7	5.0×10^3	Hayes, 1965
Aspergillus	1000	4.0×10^7	4.0×10^4	Pritchard, 1960a
Drosophila	280	8.0×10^7	3.0×10^5	Pontecorvo, 1958
Mouse	1954	5.0×10^9	3.0×10^6	

practice long distances are usually determined by the method of summation of shorter intermediate intervals (*rf* below 10%) at which all mapping functions are linear. In this case an *rf* value of 1% is taken as the unit of distance. In the Russian literature the name "morganida" is given to this unit (Lobashev, 1967), while elsewhere it is known as the "centimorgan" (Bailey, 1961) or, more customarily, as the "map unit." The distances between genes for which *rf* is greater than 10% can be roughly determined by converting *rf* into units of length with the aid of Haldane's ideal mapping function.

The problem of determining the mapping function with strict accuracy arises if it is necessary to extrapolate results obtained over a limited sector to the chromosome as a whole in order to determine its total genetic length.* This is required, for example, if the physical scale of genetic linkage maps is to be determined.

1.8. The Physical Scale of Genetic (Linkage) Maps

Despite the fact that the genetic distance, measured in map units, reflects the real physical distance between genes in the chromosome, the physical scale (expressed as the number of base pairs per map unit) or linkage maps may differ considerably in different systems (Table 1).

This is because the total intensity of recombination processes is determined by the chromosome structure and the genotype. Although rough agreement between genetic and physical distances has been clearly proved not only for higher organisms (Dobzhansky, 1929; Lefevre, 1971), but also for microorganisms (Meselson and Weigle, 1961), agreement in detail is never observed. In the centromere regions of *Drosophila* chromosomes the frequency of crossing-over is lower than in the distal regions (Mather,

---

*The genetic length of a chromosome is most easily determined by summation of the *rf* values for short segments. However, this method cannot always be used because of the insufficient density of chromosomal markers.

1933a). The linkage map is apparently reduced in scale distally. In *Podospora anserina,* on the other hand, an increase in the frequency of crossing-over has been found near the centromere (Rizet and Engelman, 1949). In the middle of the chromosome of phage T1 crossing-over is suppressed, as a result of which the genes are stretched in the center of the map although in fact the physical distribution of the genes is uniform along the whole length of the chromosome (Michalke, 1967). Disagreement between the genetic and physical lengths is found in the region of genes 34-35 of phage T4 (Childs, 1971; Beckendorf and Wilson, 1972). The total length of the heterochromatin segments of the *Drosophila* chromosomes does not correspond to their genetic length. The latter is only 0.04 map unit in the X-chromosome, while cytological determinations give much higher values: 0.5 of the X-chromosome in metaphase or 17 of the 1024 bands in the polytene chromosomes of the salivary glands of *Drosophila* larvae, which is equivalent to 33 or 1.6 map unit, since the genetic length of the X-chromosome is 66 map units (Roberts, 1965; see the survey by King, 1970). Since the frequency of crossing-over is under genetic control (3.8) and mutation in the corresponding gene can reduce the total recombination frequency by a thousand times (4.8), it is clear that the scale of the linkage maps may vary to a similar degree. The scales of the maps in the same organism may also differ by an order of magnitude if different methods of hybridization (5.6) or different alleles (2.12) are used. Only the relative distances on the map and, of course, the physical dimensions of the chromosomes and genes are invariable.

The examples given above illustrate an extremely important feature of linkage maps which must never be forgotten: their formal character. Mapping is merely a graphical method of representing experimental results. The map is not an exact copy of a chromosome but one distorted by the recombination behavior of the chromosome. A clear example of formalization is given by the circular character of the linkage map of phage T4, the chromosome of which consists of a linear open DNA molecule (Foss and Stahl, 1963; Streisinger et al., 1964; see also page 202).

1.9. Chromatid Crossing-over

According to Janssens' theory crossing-over must take place between two of the four chromatids, for each chiasma is usually formed by the crossing of only two of the threads (partial chiasmatype). However, it was considered at the same time that chromosomes divide into two only in the diplotene stage, and this stage, with its marked shortening and thickening of the chromosomes, is unfavorable for breaking and reunion to take place at precisely homologous loci. For this reason Muller (1916) suggested that crossing-over takes place at the later stages, evidently at the zygotene stage.

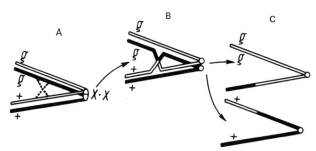

Fig. 5. Chromatid crossing-over in linked *Drosophila* X-chromosomes. A) Linked X-chromosomes (X·X) after replication. Sister chromatids shown in different colors. Dots denote position of future chiasma. Circles represent centromeres. Original X·X chromosome heterozygous with respect to gene *g*. For simplification of the diagram the ramaining markers are not shown. B) Separation of centromeres during second division. C) Two gametes homozygous with respect to alleles *g* and +. Two other gametes containing Y-chromosomes are not shown.

Consequently, crossing-over must take place at the chromosome and not at the chromatid level. The first indications that crossing-over can take place at the four-chromatid stage, i.e., after division of the chromosomes, were obtained by Bridges in 1916. However, these results could be interpreted in another way (Muller, 1916). Using an elegant genetic method, Bridges and Anderson obtained solid proof that ultimately chromatid crossing-over is a possibility (Anderson, 1925; Bridges and Anderson, 1925). In *Drosophila* females with two linked X-chromosomes (X·X) as well as a Y-chromosome, can be obtained. If such females are crossed with normal males, normal sons and daughters just like the mother (X·X) are formed (Table 2).

Having obtained the X·X females heterozygous with respect to the garnet, forked, and vermilion genes, Anderson crossed them with normal males and found daughters homozygous with respect to these markers in the progeny. He found, moreover, that the more distant the gene from the

TABLE 2. Results of Crossing X·X *Drosophila* Females with Normal Males

Gametes of females	Gametes of males	
	X	Y
X·X	X·XX (dies)	X·XY
Y	XY	YY (dies)

centromere, the higher the percentage of homozygosis relative to it (for garnet 10%, forked 5.5%, and vermilion 16.1%).

As Fig. 5 shows, homozygotes could be isolated only if crossing-over took place between the gene and centromere. Moreover, this result could be obtained only if only two of the four chromatids took part in the exchange.

Despite the evidence of these and other experiments performed later on corn trisomics by Rhoades (1933) one important reservation must be made about them: these results were obtained on exceptional (X · X or trisomics) individuals. In addition, these facts did not rule out the possibility that crossing-over may be both chromatid and chromosome. All these objections were overcome by tetrad analysis (Lindegren, 1933).

Tetrad analysis is possible only in organisms in which all the end products of individual meiosis remain viable and intact, unmixed with the others. In *Drosophila* (and in all higher animals and plants except liverworts) only one of the four haploid cells formed as the result of meiosis survives in oögenesis and the remaining polar bodies are destroyed. Because it is impossible to preserve the products of individual meiosis, recombination in these organisms can be studied only by statistical analysis of the gametes (progeny). In the lower fungi and, in particular, in *Neurospora crassa,* the four haploid cells (tetrad) formed by meiosis remain together for some time, covered by a common membrane (the ascus). The arrangement of the spores in the ascus is linear, since the spindles of the first and second divisions lie on the same axis. Meiosis is followed by postmeiotic division (mitosis), so that as a result the ascus contains four pairs of spores, which can be removed from it by means of a micromanipulator (Zakharov and Kvitko, 1967).

Let us examine the cross between two genes *ab* × +, in which the genes *a* and *b* are linked (Fig. 6). In the left half of the diagram (A) is shown the expected segregation in the asci if the crossing-over takes place at the chromosome level; in the right half (B) the same at the chromatid level. As the diagrams show, chromosome crossing-over may lead to the formation of tetrads of two types: tetrads with spores possessing the parental genotypes (P ditypes) and tetrads with recombinant genotypes (nonparental or N ditype). Clearly the first type is indistinguishable from those tetrads whose formation involved no crossing-over, so that crossing-over between the gene and centromere would be impossible to find. Exchanges between the genes would lead to the formation of N tetrads only. In fact the picture was much more complex: besides P tetrads in two-gene crosses T tetrads (tetratypes) and tetrads with an arrangement of the spores which indicated that the two nuclei formed after the first meiotic division are heterozygous, and that segregation takes place in the second division, were most frequently formed. Such an arrangement of the spores could only be obtained as the result of chromatid crossing-over between the genes and

§1.10] CHROMATID INTERFERENCE

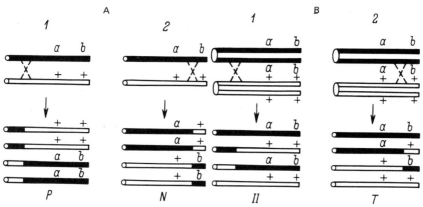

Fig. 6. Types of tetrads depending on the stage at which crossing-over takes place. A) Chromosome stage, B) chromatid stage: 1) crossing-over between markers and centromere; 2) crossing-over between markers. Homologous chromosomes in the figure are shown in a different color. Below, under the arrows, types of tetrads: P) parental ditype, N) nonparental ditype, II) segregation in second division, and T) tetratype.

centromere. N tetrads also appear, but in very low frequency. The reason for their appearance will be clear from Fig. 7 (1.10). The results of tetrad analysis thus simply and conclusively prove the chromatid basis of crossing-over.

Tetrad analysis also shows that reciprocal recombinants (R_1 and R_2) are formed during crosses between genes as a result of a single act of exchange. In other words, crossing-over is a mutual (reciprocal) exchange of information between homologous chromosomes. Reciprocal products of crossing-over have also been found in *Drosophila* on linked X-chromosomes (Beadle and Emerson, 1935). Another significant fact is the discovery that there is no limitation in the choice of pairs for crossing-over: any chromatid of one chromosome can be exchanged with any other chromatid partner.

A curious result follows from the chromatid character of crossing-over: one exchange leads to the formation of two recombinant and two parental gametes (spores). Consequently, the true frequency of crossing-over is twice as high as the value of *rf* determined in a random sample of the progeny (this statement, of course, is valid only for distances at which double exchanges can be disregarded).

1.10. The Problem of Chromatid Interference

Haldane (1931) pointed out that interference can be explained on the basis of chromatid crossing-over on the assumption that adjacent exchanges take place as a rule between different pairs of chromatids and not between

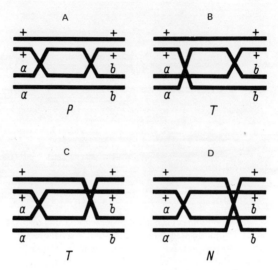

Fig. 7. Different types of double crossovers at the four-chromatid stage and types of tetrads (P, T, and N). A) Two-chromatid, B and C) three-chromatid, D) four-chromatid double exchanges.

the same pair. In other words, four-chromatid double exchanges (Fig. 7) arise more often in a short distance than two-chromatid exchanges. Since a four-chromatid double exchange does not lead to the formation of double recombinants (Fig. 7), interference arises even when the distribution of the crossing-over points along the chromosome is random in character. The distribution of exchanges is nonstatistical not along the chromosome, but "crosswise," i.e., between the chromatids taking part in the consecutive exchanges. This hypothetical phenomenon was described by Mather (1933a) as chromatid interference.

The problem of chromatid interference can also be resolved only by the aid of tetrad analysis. Let us turn once again to the analysis of segregation in the asci. Different types of double crossovers are illustrated in Fig. 7. Depending on which chromatids take part in these exchanges, different types of tetrads (P, T, and N) are formed. In the absence of chromatid interference the ratio between the two-chromatid double exchanges and three- and four-chromatid double exchanges must be 1:2:1 (Beadle and Emerson, 1935). Since in this case the number of two-chromatid double exchanges is equal to the number of four-chromatid exchanges, $rf_{max} = 50\%$. If there are more two-chromatid exchanges, $rf_{max} < 50\%$, but if there are more four-chromatid exchanges, $rf_{max} > 50\%$ (Emerson and Rhoades, 1933; Weinstein, 1958). Since in all systems so far studied a value of $rf_{max} > 50\%$

has never been observed (the exceptions found in some cases are the result of cytological anomalies; Wright, 1947), it can be concluded that chromatid interference does not take place. However, this is not a decisive argument.

Emerson (1968) generalized the results obtained by the study of chromatid interference by tetrad analysis. The results as a whole indicate absence of chromatid interference (0.29 : 0.48 : 0.23), although there is a small yet statistically significant excess of two-chromatid over four-chromatid exchanges and marked heterogeneity of the results obtained by different marker densities (the lower the density of genetic markers in a particular segment studied, the greater the error introduced into the measurements on account of unrecorded multiple exchanges). Semitetrad* analysis (on linked X-chromosomes) in *Drosophila* also reveals a very slight tendency for the number of two-chromatid exchanges to exceed that of four-chromatid (578 : 527), but this is balanced to some extent by the fact that no three-chromatid exchanges are recorded in this type of analysis (half of them appear as two-chromatid and the other as four-chromatid double crossovers).

Chromatid interference has been found only in specific chromosome regions in *N. crassa* (Lindegren and Lindegren, 1942): when simultaneous exchanges were recorded in regions on opposite sides of the centromere, more two-chromatid than four-chromatid double crossovers were found. However, more recently this excess of double exchanges has tended to be ascribed rather to the existence of structural heterozygosis (inversions, for example) than to chromatid interference proper (Stadler, 1956; Perkins, 1967). The absence of chromatid interference means that chromosome interference must be regarded as the only cause of genetic interference.

Positive chromosome interference is revealed not only by statistical, but also by tetrad analysis: it leads to the appearance of T tetrads with a frequency much higher than that which would be expected in the absence of interference, and to the virtually complete absence of N tetrads if recombination is analyzed in short segments (Perkins, 1953. Bole-Gowda et al., 1962). It is therefore because of chromosome interference that all four chromatids cannot participate simultaneously in the same act of crossing-over and that only any two chromatids (not sister chromatids) of the four can so participate. The effect of one exchange on another is therefore not confined to the same chromatids but also extends to the sister chromatids. Consequently, despite the fact that the chromosome in meiosis consists of two chromatids, from the point of view of recombination it behaves as a single entity.

---

*In semitetrad analysis only two of the four recombining chromatids are considered. This type of analysis is possible in diploid organisms during mitotic recombination (1.16 and 3.4) and also in meiosis, if one of the chromosomes in a haploid cell (gamete) is duplicated (for example, $X \cdot X$ in *Drosophila*).

1.11. Chiasmata and Crossing-over

On the basis of his observation that the number of chiasmata in bivalents of *Primula sinensis* decreases during the transition from the diplotene stage to metaphase, Wenrich (1916) postulated that crossing-over is the result of breakage of "classical" chiasmata. The best known supporter of this hypothesis in later years was Sax (1930, 1931). However, Darlington (1937) showed that the decrease in the number of chiasmata is not due to their breakage, as Wenrich thought, but to movement of the chiasmata toward the ends of the bivalent as the homologous chromosomes separate, which they begin to do in the diplotene stage. The process of "sliding" of the chiasmata toward the ends of the bivalent was described by Darlington as terminalization of the chiasmata.

Wenrich's hypothesis does not require exact agreement between the frequency of chiasmata and *rf*. According to the chiasmatype theory such agreement is essential: in short segments where there are very few double exchanges the frequency of the chiasmata must be twice the *rf* value (1.9). Beadle (1932) studied the relationship between these values in *Zea mays*. In experiments to study translocation between the 8th and 9th chromosomes he showed that the frequency of chiasmata in a region 12 map units long is 20%. Darlington (1934a) determined the total number of chiasmata for all *Zea mays* chromosomes, converted the numbers into map units (1 chiasma = 50 map units), distributed it among 10 chromosomes on the basis of their length in the pachytene stage, and thus obtained an approximate value for the total genetic length of each chromosome. When he compared the genetic lengths of the chromosomes thus obtained with the values observed he found that no chromosome has an observed length greater than that expected theoretically. Since the observed lengths are minimal (because of incomplete saturation of the map with markers) these results provide solid confirmation for the chiasmatype theory (see also Rhoades, 1950).

It follows from the chiasmatype theory that the conjugated chromatids on both sides of the chiasma are sisters (Fig. 2). The classical theory, on the other hand, postulates that during chiasma formation there is an exchange of partners and that pairing between sisters on one side of the chiasma gives way to pairing not between sisters. Cytological observations on the behavior of unequal chromosome pairs, polyploids, and bivalents heterozygous for inversions in meiosis have shown that the chiasmatype theory also holds good in this respect (Darlington, 1937; Callan and Lloyd, 1956; Zen, 1964).

The chiasmatype theory postulates that chiasmata are the result of a previous crossing-over. If this is true and if genetic interference is the result of chromosome rather than of chromatid interference, it must be manifested as interference of chiasmata. Haldane (1931), after statistical analysis of extensive cytological material obtained with *Vicia faba* (Maeda, 1930), showed that chiasma interference does in fact exist.

Direct comparison of the frequency of cytologically observed double chiasmata in paracentric* inversion in *Lilium formosanum* with the frequency of two-, three-, and four-chromatid double exchanges, leading to the formation of bridges and fragments in anaphase, showed unequivocal agreement between chiasmata and crossing-over (Brown and Zohary, 1955). Moreover, the relative frequencies of multichromatid exchanges corresponded to those expected in the absence of chromatid interference. These observations gave powerful support to Janssens' theory (see also Brandham, 1969).

The main objection against the chiasmatype theory for a long time was the fact that chiasmata are found in *Drosophila* males, in which crossing-over never takes place. Chiasmata in males were reported by Darlington (1934b) and later by Cooper (1949), but it has since been shown that these observations were mistaken. The artefact could have arisen because of the exceptionally unfavorable nature of this material for cytological observation (Slizynski, 1964). The absence of the typical picture of the synaptinemal complex, or synapton (4.6), in males has also been demonstrated with the electron microscope (Meyer, 1960). Achiasmatic meiosis also correlates with the absence of a synapton in other organisms (see 4.6).

Decisive evidence in support of Janssens' theory was obtained when it was shown that lines of achiasmatics can be obtained in which mutation in one gene leads to the suppression of crossing-over and to the simultaneous disappearance of chiasmata and of the synapton (Enns and Larter, 1963; Smith and King, 1968).

1.12. Cytological Evidence for Crossing-over

Creighton and McClintock (1931), using suitable strains of *Zea mays*, obtained hybrids in which the 9th pair of chromosomes was heteromorphic (Fig. 8, P_1): one was normal but the other was longer as the result of reciprocal translocation of the material of the 8th chromosome into one of its arms. In addition, this same chromosome had an additional cytological marker: a knob at the end of the other arm. The normal chromosome contained the genes c (colorless endosperm) and Wx (starchy endosperm), lying in its short arm. In the heterozygote the alleles C and wx were found. These diheterozygous (genetically and cytologically) plants were crossed with a strain (P_2) with morphologically normal chromosomes, homozygous for c and wx:

$$\frac{Cwx}{cWx} \times \frac{cwx}{cwx}$$

---

*Inversion in a chromosome arm not affecting the centromere.

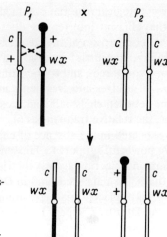

Fig. 8. Cytological evidence for crossing-over in corn (schematic). Filled circles denote cytological marker (nodule), empty circles denote centromere. Cytogenetic structure of recombinants is shown below.

Recombinant (CWx and cwx) grains were found in the progeny and cytological analysis was carried out to discover whether chromosome material had been exchanged in the segment between the cytological markers. It was found that genetic recombinations in the region bounded by the heteromorphic loci were invariably accompanied by exchange of the segments between the two chromosomes: the recombinants contained chromosome 9, which was either of the normal length but with a knob (R_2) or without a knob but elongated (R_1). The coincidence of the cytological and genetic crossing-over was striking. A similar correlation was found for *Drosophila* (Stern, 1931) and later for bacteriophage λ (4.7).

1.13. Darlington's Hypothesis

Darlington, who was responsible for much of the work in support of the chiasmatype theory, developed over a period of years (Darlington, 1937) a hypothesis of mechanical crossing-over based on the idea (Bridges, 1915) of physical breakages and reunions of the chromosomes due to coiling and stretching. Darlington postulated that the forces keeping sister chromatids side by side in the prophase of mitosis are also responsible for zygotene pairing of homologous chromosomes, for at that moment the chromosomes are still undivided and conjugation satisfies their pairing urge (the precocity theory). During replication of the chromosomes (which in his opinion occurs in the diplotene stage), now that their pairing urge has begun to be satisfied by the newly formed sister chromatids, the attraction between the homologues weakens and the chromosomes begin to repel each other. Since the chromosomes at this stage of meiosis are intercoiled, the repulsion forces cause stretching in the chromatids, leading to rupture of one of them. As a result equilibrium in the intercoiling of the homologues is disturbed. To restore the equilibrium the degree of internal coiling of the homologous chromosome is increased as much as possible at the homologous locus, leading to breakage of one of the chromatids in this chromosome also. Rotation of the ends of the broken chromatids around the intact sister strands as the result of their residual coiling leads to contact and to reunion.

1.14. Belling's Hypothesis

For many years Janssens' theory was criticized by most cytologists. One of the arguments put forward by the opponents of his theory was that in the early diplotene stage no evidence of breakage and reunion can be found cytologically. Because of this criticism and on the basis of other observations on meiosis in *Lilium* Belling (1928) postulated that exchanges between chromatids take place at the pachytene stage during conjugation of the chromosomes. Exchanges at this stage cannot be found directly, and their aftereffects (chiasmata) become visible much later. This modification marked a significant contribution to the development of the chiasmatype theory.

Later, in rejecting Darlington's coiling theory because it could not explain why the breaks resulting from stretching appear at strictly homologous points, Belling (1933) suggested an alternative mechanism of crossing-over (Fig. 9). He postulated that during division of the chromosomes which, as was thought at that time, consist of chains of chromomeres, individual chromomeres reproduce first, and they subsequently are joined by longitudinal bonds (chromonemes), whose replication takes place somewhat later. Before the appearance of the longitudinal bonds, the chromomeres remain free for some time and they are kept in linear order by attraction toward the homologous chromomeres of the maternal chromatid, which still remains intact. Since the chromosomes are coiled and the new bonds are formed along the shortest path, reunion and crossing-over are possible.

Belling's hypothesis met with one obvious difficulty, namely that the possibility of crossing-over is restricted to the two newly synthesized chromatids (negative chromatid interference). To reconcile the hypothesis with the genetic evidence that all four chromatids can take part in exchanges, Belling was forced to postulate additional exchanges between sister chromatids (1.15). However, he was not convinced of the existence of such exchanges. He therefore suggested that during replication the chromomeres are not fixed to the old chromoneme, but "hang" with equal probability between the old and new chromonemes. Under these circumstances, dif-

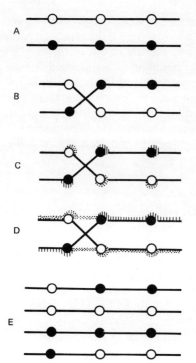

Fig. 9. The mechanism of crossing-over as conceived by Belling (1933). Circles denote chromomeres; lines, chromonemes; dots and shading, replicated material. A) Original position of two homologous, unreplicated chromosomes; B) coiling of homologues; C) replication of chromomeres; D) formation of new bonds between chromomeres, leading to crossing-over; E) tetrad.

ferent patterns of intercoiling of the chromosomes enable the formation of three-and four-chromatid double crossovers. In molecular terms, this hypothesis was recently revived by Uhl (1965), who suggested that crossing-over takes place by the incorrect attachment of protein linkage-groups (4.2) to DNA molecules at the time of DNA replication.

1.15. Sister-Strand Exchanges

Belling's hypothesis in its initial form, including sister exchanges, was supported by the Lindegrens (Lindegren and Lindegren, 1937; Lindegren, 1964). The scheme in Fig. 10 illustrates how sister-strand crossing-over can make three- and four-chromatid double exchanges possible. Exchange between sisters in the region between two chiasmata leads to three-chromatid double exchange (Fig. 10A), while sister-strand exchanges in each of the homologues leads to four-chromatid double exchange (Fig. 10B). If these changes are frequent enough (so that their even and odd numbers in this region are equal), four classes of double crossovers will appear at the frequency required in the absence of chromatid interference, namely 1 : 1 : 1 : 1 (1.10).

It is difficult to demonstrate the presence of exchanges between sister chromatids, if such exist, for these chromatids are genetically identical. During unequal crossing-over, when it would be possible to detect such exchanges, Sturtevant (1925) failed to find any. Schwartz (1953), who studied the effects of crossing-over in *Zea mays* heterozygous for the 6th normal and 6th circular chromosomes, found a high frequency of single and double bridges in the second anaphase, which could have been due either to negative chromatid interference or to the presence of sister-strand exchanges. Having decided in favor of the second hypothesis, Schwartz calculated that the number of exchanges between sister chromatids to one bivalent is very high.

Direct proof of the presence of sister-strand crossing-over in mitosis was obtained by Taylor (survey: Taylor, 1964) using the autoradiographic method. However, according to their findings the frequency of the exchanges is low (on the average one exchange per bivalent). A special investigation of sister-strand exchanges in meiosis showed that they place only in premeiotic stages (Taylor, 1967).

The experiments on *Drosophila* carried out by Schwartz (1954) to obtain evidence in support of his hypothesis were technically faulty (Baker and Swatek, 1965).

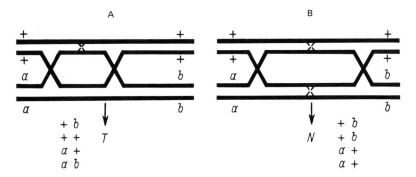

Fig. 10. Conversion of two-chromatid double crossing-over into three-chromatid double exchange in the presence of one additional sister-strand exchange (A) and into four-chromatid double exchange in the presence of two such crossovers (B). Below: tetrads, T and N. Broken lines denote sister-strand exchanges.

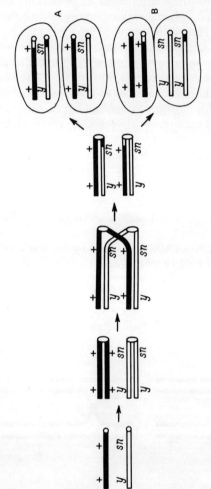

Fig. 11. Mitotic crossing-over in a diheterozygous diploid cell. A) Complementary products of recombination between markers and the centromere move toward the same pole and the cell remains heterozygous; B) recombinant chromosomes move apart toward different poles, giving rise to homozygous daughter cells.

1.16. Mitotic Crossing-over

Stern (1936) pointed out that in some *Drosophila* females diheterozygous for the y and sn genes (y determining yellow body color, sn singed bristles) spots corresponding to recessive characters appear on the body. Stern attributed the formation of these spots to the presence of mitotic crossing-over at the four-strand stage (Fig. 11). The so-called twin spots can appear only if crossing-over takes place between the sn gene and the centromere and if in anaphase the y^+sn^+ and y^+sn^+ chromosomes move away to one pole and the $y\,sn$ and $y\,sn$ chromosomes to the other. Mitotic crossing-over takes place in both males and females. Although mitotic crossing-over was discovered in *Drosophila,* the most suitable objects for its study are fungi (Pontecorvo and Roper, 1953). Mitotic crossing-over is a rare event, taking place thousands of times less frequently than meiotic crossing-over. Moreover, even if crossing-over does takes place, it is rarely anything more than one or two exchanges in the whole complement of chromosomes. Crossing-over takes place at the four chromatid stage (however, see Wildenberg, 1970) and is characterized by reciprocity. The frequency of mitotic recombinations can be increased by the action of ultraviolet light, x-rays, inhibitors of DNA synthesis, and chemical mutagens (different aspects of mitotic crossing-over are discussed by Pontecorvo and Cafer, 1958; Roper, 1966; Holliday, 1968; Nakai and Mortimer, 1969; Zimmermann, 1971; see also 4.10).

1.17. Merozygote Systems

Both meiotic and mitotic crossovers take place in diploid cells, i.e., whole and equal chromosomes of each parent participate to an equal degree in the act of recombination. The term merozygote (Jacob and Wolman, 1962) indicates that the recombination takes place not in completely diploid, but in only partially diploid (merodiploid) cells. The parents make unequal contributions: one (the recipient) a whole chromosome, the other (donor) only part of a chromosome. The whole of bacterial genetics is based on the use of merozygote systems. During conjugation there is a gradual transfer of the chromosome of the male (Hfr) cell into the female (F⁻); this process can be stopped at any moment and only as much of the male chromosome as has succeeded in penetrating by that time will be found in the female cell. Under different conditions of mixing of the genetic material of different cells (transduction, sexduction, and transformation) the size of the transferred fragments is limited either by the size of the phages (transduction) or by the size of the sex factor (sexduction), or by fragmentation of the DNA during its liberation from the donor cells (transformation).

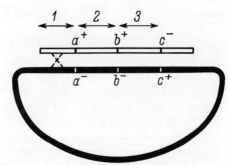

Fig. 12. Three-marker crossing in a merozygote system: 1) interval between end of donor fragment (exogenote) and marker a; 2 and 3) intervals between markers. Exogenote colored white, circular chromosome of recipient cell colored black. Imagine what happens if the crossing-over takes place only in the region marked by the broken lines.

Because of the one-way character of the transfer of genetic material from cell to cell and the formation of incomplete zygotes, limitations are imposed on the recombination process. As Fig. 12 shows, for the recombinant chromosome to preserve its structure, there must be an even number of exchanges in the merozygote. If the number is odd, an open ring will be formed with terminal duplications. Consequently, the elementary act of recombination which is recordable in the merozygote is a double crossing-over (single exchanges in bacteria nevertheless sometimes take place in cases of specialized recombination such as lysogenization of phages and the conversion of F^+ cells into Hfr). Double exchange must result in reciprocal integration of the donor fragment into the recipient chromosome and of a segment of the recipient chromosome into the donor fragment. Since the donor fragment is usually not viable, only integration into the chromosome can be detected, and for this reason it has only very recently become known that integration is in fact reciprocal.

To shed light on this problem, a donor fragment replicating spontaneously in the cytoplasm was used (Herman, 1965; Meselson, 1967). It is able to replicate because it is an integral part of the episome of *Escherichia coli* — it is linked with the sex factor (F'). As the fragment to be transferred, Meselson used the chromosome of prophage λ. The merodiploid which he constructed thus had two differently marked prophages (λ^+ and $\lambda c\,mi\,h$), one of which occupied its usual place in the chromosome, while the other was in the F' episome. During reproduction of these merodiploids prophase λ is spontaneously induced with low frequency simultaneously from the episome and from the chromosome; it propagates in the cell and causes its lysis, and escapes into the medium. The culture therefore always contains free phage. Interaction between homologous segments of the episome and chromosome an d recombination take place in the cells with a definite probability. The two recombinant products remain in the cell and can be found after induction.

For the experiment 5735 single cells were selected from a culture of the merodiploid strain and each was propagated separately (~30 generations). These individual cultures were analyzed for their content of free phage particles. Only phages of parent types ($\lambda c\, mi\, h$ and $\lambda+$) were found in equal numbers in 5136 cultures, 437 cultures contained recombinant phages, and only phages of one genotype (homozygotes) were found in 162 cultures. Most cultures containing recombinant phages contained both reciprocal (complementary) types (for example, $\lambda+ mi\, h$ and $\lambda c++$). However, nonreciprocal recombinants were found in low frequency in one culture ($\lambda c++$ and $\lambda++h$). To explain the origin of the nonreciprocal recombinants and homozygous cultures Meselson postulated that recombination in merozygote systems takes place at the four-chromatid stage. Then, just as in mitotic crossing-over, the absence of reciprocal recombination products means that they segregate at random into different daughter cells (see Fig. 41). However, these rare cases are possibly due to correction.

How is rf determined in merozygote systems? A special feature of these systems is that we do not know the original number of zygotes, and this makes it difficult to establish the normal probability of recombination. The only approach in this case is to measure the fraction of recombinants of a particular type relative to another recombinant fraction. Let us consider the three-marker cross $a^+b^+c^- \to a^-b^-c^+$ (the arrow denotes the direction of transfer of the genetic material). In an ordinary three-marker cross all 8 classes of progeny can be analyzed. In the merozygote system we select the a^+c^+ recombinants and then determine the frequency of appearance of unselected b and b^+ alleles in them. As Fig. 12 shows, $a^+b^+c^+$ recombinants appear if recombination takes place in segments 1 and 3 but not in segment 2: $rf_1 rf_3 (1-rf_2)$. The expected frequencies of appearance of recombinants of either class, however, can be determined only if Z (the original number of zygotes) is known. Since this value is unknown, it is impossible to determine the absolute values of rf. All that can be stated is that if recombinants of one class (for example, $a^+b^-c^+$) are found with a frequency x times less (or more) than recombinants of another class ($a^+b^+c^+$), rf_2 is x times less (or greater) than rf_3. Even this assertion clearly is not sufficiently strict. In order to progress from relative to absolute values, additional data are required. A method of determining the absolute value of rf in merozygote systems was suggested by Lederberg (1947) and rendered in exact mathematical form by Bailey (1951).

Let us consider the four-marker cross $a^+b^+c^+d^- \to a^-b^-c^-d^+$ (Fig. 13), in which a^+d^+ recombinants are selected. By analogy with the preceding

Fig. 13. Four-marker cross in a merozygote system: 1, 2, 3) intervals between markers.

TABLE 3. Expected and Observed Frequencies of Appearance of Recombinants in a Four-Marker Cross in a Merozygote System (Bailey, 1951)

Crossing over in interval	Expected frequency	Observed frequency	Genotype
1	$Zrf_1(1-rf_2)(1-rf_3)/R$	a	$a^+b^-c^-d^+$
2	$Z(1-rf_1)rf_2(1-rf_3)/R$	b	$a^+b^+c^-d^+$
3	$Z(1-rf_1)(1-rf_2)rf_3/R$	c	$a^+b^+c^+d^+$
1, 2, 3	$Zrf_1 \cdot rf_2 \cdot rf_3/R$	d	$a^+b^-c^+d^+$

Note. R denotes recombinants as a fraction of the whole interval: $R = rf_1(1-rf_2)(1-rf_3) + (1-rf_1)rf_2(1-rf_3) + (1-rf_1)(1-rf_2)rf_3 + rf_1 rf_2 rf_3$.

reasoning for the three-marker cross, we can determine the expected and observed frequencies of appearance of recombinants of the four classes accessible for observation (Table 3).

Since the number of equations is equal to the number of unknowns, and since the unknown number of zygotes Z is excluded as the result of standardization, the values of rf_1, rf_2, and rf_3 can be determined separately:

$$\frac{rf_1}{1-rf_1} = \left(\frac{ad}{bc}\right)^{1/2}, \quad \frac{rf_2}{1-rf_2} = \left(\frac{bd}{ac}\right)^{1/2}, \quad \frac{rf_3}{1-rf_3} = \left(\frac{cd}{ab}\right)^{1/2}.$$

The values of rf obtained in this way are absolute and can be converted into values of d (distance) with the help of Haldane's ideal mapping function. The use of this function takes for granted the absence of chromosome interference. The mathematical theory of recombination at great distances (during conjugation), based on this assumption, gives results in good agreement with those obtained experimentally (Verhoef and de Haan, 1966; Wu, 1967). However, it would be premature to draw a categorical conclusion regarding the absence of chromosome interference during conjugation for no systematic research in this direction has yet been undertaken.*

---

*In some cases (Maccacaro and Hayes, 1961) negative interference extending over a distance of up to 100 map units is found during conjugation. It is caused by an increase in the frequency of recombinations at the ends of the exogenote (Wood, 1969).

If certain conditions are satisfied in a conjugation experiment absolute values of rf can thus be obtained directly in two-marker (ignoring the selecting marker) crosses. These conditions are: the strains crossed must be isogenic (so that any one allele is not preferentially incorporated into the recombinant chromosome) and the ends of the transferred fragment must be large enough for recombination equilibrium to have been established in them (an equal number of even and odd exchanges). This last condition is not essential, however, for there is increasing evidence to suggest that this equilibrium is established at the ends regardless of their dimensions (Pittard and Walker, 1967; Mosig, 1967; Doermann and Parma, 1967). Hence, in a cross of the $a^+b^+ \to a^-b^-$ type:

$$rf_{ab} = \frac{a^+b^-}{a^+b^+ + a^+b^-} = \frac{rf_1 rf_2 (1 - rf_3)}{rf_1 (1 - rf_2) rf_3 + rf_1 rf_2 (1 - rf_3)},$$

where rf_1 and rf_3 are the recombination frequencies at the ends of the fragment. Substituting in this equation $rf_1 = rf_3 = 0.5$, we obtain $rf_{ab} = rf_2$. It thus also follows that rf cannot exceed 50% during normal conjugation. Merozygote systems in which the recombination frequency can exceed this value (transformation and transduction) will be examined in 5.6.

1.18. Recombination in Bacteriophages

Recombination in bacteriophages takes place during mixed infection of bacterial cells by genetically different phage particles. After their penetration into the cell the DNAs of both particles begin to replicate and to form a single pool of chromosomes of the vegetative phage. Within this pool intact phage genomes mate in pairs, randomly, and repeatedly. Recombination in phages is thus a problem in population genetics: the population is the pool of chromosomes of vegetative phage, numbering from tens to thousands of phage genomes.

Visconti and Delbrück (1953) put forward a quantitative theory of genetic recombination in phages. Let m denote the mean number of crosses of each chromosome in the pool during development and let p denote the frequency of appearance of recombinants in one cross; the mean number of crosses completed by recombination of linked markers will thus be mp. If the number of crosses per chromosome obeys the Poisson distribution, the probability that any (chosen at random from the pool) chromosome participates at least once in a cross leading to exchange of markers will be $1 - e^{-mp}$. If the probability that a given chromosome carries the marker of one of the parents is designated by the letter k (k is equal to the fraction of parental phages of this type during infection). The probability that its partner in the last cross, leading to recombination of the markers, will be a chromosome carrying this marker is $1 - k$. Hence, the mean frequency of

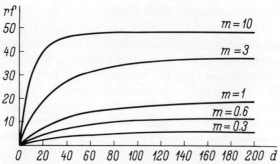

Fig. 14. Mapping functions for phages with open maps (after Stahl, 1966); m) number of rounds of mating in pool of vegetative phage; d) in map units, rf) in percent.

one of the two reciprocal recombinants in the stock is given by $k(1-k) \cdot (1-e^{-mp})$. It thus follows that

$$rf = 2k(1-k)(1-e^{-mp}).$$

If the relationship between p and the mean number of exchanges per cross (d) obeys Haldane's law, the equation obtained can be rewritten in the form

$$rf = 2k(1-k)\left\{1 - \exp\left[-\frac{m}{2}(1-e^{-2d})\right]\right\}.$$

In Fig. 14 rf is shown as a function of d for several values of m (mating in which $k_1 = k_2$).* As the figure shows, $rf_{\max}$ depends on the number of rounds of mating (m) of the phage in the pool. However, the gradient of the mapping curves also depends on the degree of chromosome interference (1.7). Since it is virtually impossible to distinguish between these two factors in phages (Steinberg and Stahl, 1967) our knowledge of chromosome interference in phages is nil. All that these mapping functions can do is to eliminate low negative interference, the cause of which is inherent in the population character of phage genetics (Doermann and Hill, 1953). There are three reasons for the excess of double recombinations: 1) the participation of single recombinants in further mating events, 2) the formation of recombinants only in some matings, whereas their frequency is calculated in terms of the entire progeny of the cross (for further details, see Hayes, 1965), and 3) the presence of paired exchanges taking place at low frequency, but inevitably (as a result of the ring-shaped character of the chromosome). One of the last

---

*These functions are valid only for phages with open linear maps. For phages with circular maps the mapping functions are more complex (Stahl and Steinberg, 1964).

two factors (or both together) lead to values of $C > 1000$ for phages 186, P2, and S13 of *E. coli* (Mandel and Kornreich, 1972). During interaction between incomplete phage chromosomes and analysis of recombination at the ends of complete chromosomes, interference is absent in the yields of individual cells (Mosig et al., 1971).

A unique feature of recombination between genes in phages is its non-reciprocal character. Correlation between the frequencies of appearance of reciprocal recombinants in the phage harvest from one cell is at a very low level or absent altogether (Hershey and Rotman, 1949; Bresch, 1955). Until recently doubts were being expressed whether reciprocity in phages might not be an artefact due to differences in the rate of maturation and liberation of the particles from the pool.

By an elegant experiment Weil (1969) dispelled these doubts. Recombination of vegetative phage λ is controlled by three independent systems: the Rec-system of bacterial cells, the recombinase system of the phage (Red-system), and the site-specific system (Int) responsible for integrating the phage into the bacterial chromosome. The Int-system is more effective than the Red-system and, in addition, it controls recombination only in a strictly defined region of the phage chromosome (between markers *b2* and *cIII*).*
In λ*susA+cIII* × λ+*b2*+ matings the progeny in harvests from single cells (Rec⁻) was analyzed. Correlation between the reciprocal recombinants was very low ($r = 0.16$) in the region *susA*−*b2* and very high ($r = 0.6$) in the region *b2*−*cIII*.

A more refined statistical analysis of these data (Kayajanian, 1972) showed that for the first region $r = 0.46-0.50$ and for the second $r = 0.72-0.83$ (for details, see 5.6).

1.19. The Copy-Choice Hypothesis

With the spread of genetic research to phage and bacterial systems the search for new ideas to explain a number of unexpected facts was intensified. In order to explain nonreciprocity in phages (1.18), for instance, Hershey (1952), developing Sturtevant's thinking, developed the "partial replica" hypothesis, the first and simplest form of copy-choice hypothesis. According to this hypothesis recombination is due to an exchange, not of material, but only of information between the chromosomes: part of the genetic information during replica synthesis is supplied by one parent, another part by the other parent. In other words the replica is switched from one template to another (Fig. 15A). Lederberg (1955), finding it difficult to

---

*For this reason the distance between markers *b2* and *cIII* is several times greater than follows from the physical scale of the map. The high efficiency of the Int-system is also explained by the presence of a long "black region" on the genetic map of phage P2 (Lindahl, 1969).

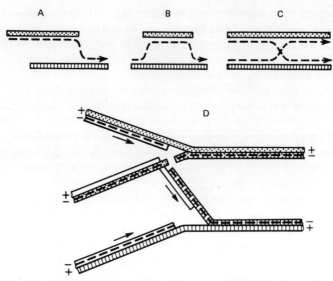

Fig. 15. Various schemes to represent the mechanism of copy-choice: A) formation of a recombinant chromosome according to the partial replica hypothesis; B) copy-choice in a merozygous system; C) reciprocal switching of the replicas giving rise to crossing-over; D) breakage and copying. A, B, C: continuous lines represent double helical DNA molecules, broken lines their "conservative" replicas. D: DNA strands of opposite polarity (+ or −) are represented by continuous and broken lines. Direction of replication is shown by arrows.

understand how double exchanges in merozygote systems take place at such a high frequency and also why they take place with such molecular precision, put forward another modification of Belling's hypothesis, the copy-choice hypothesis itself. Recombination leading to integration of the donor fragment or part of it into the chromosome is the result of a random change of templates during replication of the recipient chromosome or fragment of the donor chromosomes (Fig. 15B). Hershey and Lederberg suggested that this mechanism can only be nonreciprocal. However it was soon found that this idea, originated by Belling but described in modern terms, was so plastic that any number of modifications are possible. Freese (1957) postulated that the switching of the replicas may be reciprocal in character and, consequently, that it can explain crossing-over in higher forms also (Fig. 15C). However, this called for a further assumption regarding the conservative (and not semiconservative) synthesis of DNA (or of the chromosome): reciprocal switches of replicas during semiconservative replication would in fact lead to chromosome crossing-over. Admittedly reciprocal replica switching could take place with only one of the complementary strands of each homologous molecule. In this case, however, the chromatids would be hybrid over long segments. This situation does not arise either in classical or in bacterial systems (the purity of pairwise crossed spores in octads of fungi). In this case structural hindrances also arise, so that breakage of the parental DNA molecules had to be introduced into the scheme (Fig. 15D): the mechanism known as "breakage and copying" (Delbrück and Stent, 1957; Boon and Zinder, 1971).

The genetic and biochemical data (Chapter 4) as a whole now show that the mechanism of copy-choice does not take place as such either in eukaryotes or in prokaryotes. Its individual elements, nevertheless, are evidently found in the chain of enzyme reactions which go to form the crossing-over process (4.9).

1.20. Conclusion

Recombination of linked genes is based on crossing-over, a process of exchange of genetic information between two homologous chromosomes, resulting in chiasma formation in eukaryotes. Crossing-over usually takes place in meiosis, but it may also occur extremely rarely in mitotically dividing cells.

Crossing-over in eukaryotes has the following principal features:

1. It takes place at the four-chromatid level, but only two nonsister chromatids participate in each act of exchange, so that two crossed and two uncrossed chromatids are formed. Exchanges of sister chromatids take place extremely infrequently and, evidently, only in mitosis.
2. The two exchanged chromatids formed by crossing-over are complementary to one another.
3. As a rule crossing-over is accompanied by positive interference which is based on chromosome interference or chiasma interference. Chromatid interference does not take place.
4. The recombination frequency (rf) is a function of genetic distance (d), expressed as the frequency of crossing-over. To convert rf into d, a mapping function must be assigned, for direct equality of the two values is impossible on account of the existence of multiple exchanges and interference. The parameter d is a linear function of physical distance. However, in each concrete segment of the chromosome the frequency of crossing-over is determined by its local structure (for example, in heterochromatin regions the frequency of crossing-over is lower than in euchromatin regions). In addition, the total intensity of recombination processes is under genetic control; for these reasons the physical scale of genetic maps can differ sharply in different organisms.

The special features of crossing-over in prokaryotes are as follows:

1. Recombination in bacteriophages, taking place during vegetative development, is nonreciprocal in character. Temperate phages possess a site-specific recombination system which brings about reciprocal crossing-over.
2. In merozygote systems the elementary recordable act of genetic recombination is a double crossing-over. The original number of zygotes also is unknown. For this reason, in order to determine the absolute values of *rf* a normal must first be established with the aid of of complex recombination events. Reciprocity of recombination is observed in special cases in which the donor fragment does not die after exchange.

Nothing is yet known about the laws controlling chromosome interference in these systems. At the same time, recombination of genes in bacteriophages and, in certain cases, during conjugation also is accompanied by negative interference. The reason for its appearance is the population character of phage genetics and the fact that exchanges at the ends of the exogenote during conjugation are taken into account.

To explain the mechanism of crossing-over two fundamental hypotheses have been suggested: breakage-and-reunion and copy-choice. The first hypothesis was dominant in genetics until the mid-1950s. The study of bacterial systems and of recombination in phages has revealed many new facts which it proved impossible to explain within the framework of this hypothesis. The copy-choice hypothesis then gained in popularity. Various attempts have been made to revise the classical ideas as a result of the discovery of recombination within the gene.

CHAPTER 2

Intragenic Recombination. Random-Sample Analysis

2.1. The Concept of the Gene

It seemed obvious to Morgan and his collaborators that the gene is an indivisible unit of transmission of genetic information — a recombination unit. The gene was regarded as a discrete structure (molecule) in the chromosome capable of separation from other genes by crossing-over. The idea never even occurred that crossing-over can take place within the gene. Paradoxical as this may seem, the first indications that the gene may have a complex intragenic structure, which followed the discovery of multiple allelism strengthened this point of view still further. The allele was initially regarded as one of the two alternative states of the gene. It was soon discovered (Cuénot, 1904), however, that the gene determining coat color in the albino mouse can exist in several allelic states with different phenotypic expressions. The existence of similar series of multiple alleles has been demonstrated for many *Drosophila* genes. For example, the series of alleles of the white gene contains, besides the normal allele (w^+) determining the red color of the eyes, the mutant alleles blood (w^b), cherry (w^{ch}), eosin (w^e), apricot (w^a), and white (w). If flies carrying different alleles of the same series mated, the progeny (F_1) was found to be mutant or, in other words, the alleles belonging to the same series were noncomplementary. It was accordingly concluded that the function affected by mutation to any of the allelic states was the same in each case and, consequently, that the gene is the unit of function.

The complementation test was thus adopted as a method of gene determination: to decide whether two independently obtained mutations affecting the same character are allelic, it is simply necessary to determine the phenotypic effect of their combination in the heterozygote. If these muta-

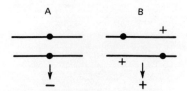

Fig. 16. Function test for allelism. Black circles represent mutant alleles. A) Allelic mutations, phenotype of the compound is mutant (−), B) mutations nonallelic, phenotype of the diheterozygote is normal (+).

tions are allelic (the heterozygote with two allelic mutations is called a compound), the phenotype of the compound will be mutant (Fig. 16). If, however, they are mutations of different genes, the phenotype of the diheterozygous individuals will be normal, since each mutant gene in the homologous chromosome has an intact (normal) allele. It will be clear why the function test of allelism is applicable only to recessive mutations. Let us assume that the function test has indicated that mutations are nonallelic. To make absolutely sure, an analytical crossing is carried out. If the genes are not linked, in 50% of the progeny wild-type recombinants will be found; if they are linked the percentage of recombinants will be lower, although they will still appear with an appreciable frequency. Meanwhile, when compounds were crossed with recessive parents, no segregation of wild-type recombinants could be observed. Consequently, the recombination test also confirmed that these mutations are allelic.

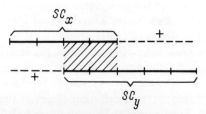

Fig. 17. Diagram showing step allelism. The mutation sc_x affects several subgenes in the basigene, mutation sc_y also affects several subgenes, but only two of them are identical with the subgenes modified by the sc_x mutation. Accordingly, when sc_x and sc_y occur together in the compound, partial complementation is observed (the noncomplementing subgenes are shaded).

2.2. Step Allelism

The first idea that the gene may have a complex internal structure was expressed at the end of the 1920s by a group of Soviet geneticists (Agol, 1929; Dubinin, 1929; Serebrovskii and Dubinin, 1929). They investigated a series of alleles of the achaete-scute gene, which determines the number and character of distribution of the bristles in *Drosophila*. Each scute allele was characterized by a definite phenotypic effect, namely as the specific character of distribution of bristles on the fly's body. Some alleles possessed an action of overlapping character. For example, the sc_3 allele caused reduction of bristles over almost the whole of the fly's body, while the action of sc_2 was limited to certain areas of the body. A common area of action was the abdomen, on which both alleles in the homozygous state determined absence of bristles. In the compound sc_3/sc_2 an unexpected effect of complementation was observed: bristles developed normally over the whole of the fly's body except the abdomen — the region of overlapping action of the alleles. On the basis of results of this type, all the *sc* alleles could be arranged, depending on the character of partial complementation in the compounds, in a certain linear order (Fig. 17), and the basic *sc* gene (what Serebrovskii called the basigene) was subdivided into several subgenes (*trans* genes). The functional integrity of the gene was in fact disturbed, for a certain degree of functional independence was ascribed to each subgene.

Further investigation showed that most of the *sc* mutations were associated with structural changes in the chromosomes whose ends were in the immediate vicinity of the *sc* gene. The question therefore arose whether the phenotypic changes obtained are the result of gene mutations or of a position effect. Unfortunately, it has not yet been possible to decide between these two possibilities (Raffel and Muller, 1940).

2.3. Pseudoallelism

Serebrovskii's group were well aware of the need for recombination analysis and that it could be used to study the internal structure of the gene. However, in Agol's words, "Unfortunately we could not use this method in our investigations because the extremely small distances between the individual areas made it impossible, at least in practice, to obtain crossing-over between them" (Agol, 1929, p. 93). In fact, to find one crossover between the yellow and achaete genes and four crossovers between achaete and scute, Dubinin and his collaborators had to examine 75,000 flies (Dubinin et al., 1937).

However, despite these ideas which were being developed by Soviet geneticists, the recombination criterion of allelism continued to hold sway in genetics. As a result, after the discovery of recombination between alleles (Oliver, 1940) geneticists still continued to put their trust in the recombination rather than the functional criterion. In a study of interaction between two alleles of the lozenge gene (determining the color and shape of the eyes) in *Drosophila*, Oliver found that reversions to the wild type, accompanied by recombination between markers linked with the lozenge locus, appeared in the progeny from analytical crossing of the compound lz^S/lz^g at a constant frequency (1×10^{-3}). By the character of distribution of these (flank) markers (2.8) the mutual arrangement of the alleles relative to the whole linkage group could be determined. By 1949 a map showing about 20 alleles of this gene had been drawn (Green and Green, 1949). Oliver was not able to determine whether classes of recombinants reciprocal to the wild type, carrying two mutant alleles on one chromosome (the *cis* configuration), appear during crossing-over between *lz* alleles, but he did observe a few cases of the appearance of phenotypically mutant flies which could have had this genotype.

Lewis (1941) conducted a detailed recombination analysis with two alleles of another gene, also determining eye development in *Drosophila*. Lewis was not aiming at discovering intragenic recombination; moreover, his investigation was partly motivated by the well-marked phenotypic development of two alleles (Star and asteroid), which could indicate that they belonged to different genes. He found that recombination takes place between them (0.02%), and later (Lewis, 1945) he found a definite phenotypic difference between the *cis* and *trans* configuration of the alleles, subsequently called the *cis–trans* effect.

In its general form the essence of this effect is that two allelic mutations, which are not complementary in a *trans* compound (the alleles are in different chromosomes), give the wild type in the *cis* compound (in the same chromosome). The fundamental importance of this effect was that, and only that a reciprocal recombination product (the *cis* compound) was found, and it could accordingly be considered that "reversions" to the wild type are actually the results of crossing-over between alleles. This meant that crossing-over can take place inside the gene. Besides the purely psychological difficulties due to the need for rejecting the recombination criterion of allelism if this fact were accepted, the then existing views of the gene as a protein molecule likewise provided no real basis for the concept of its divisibility. The alleles between which recombination had been demonstrated, and more and more of them were continually being found (survey: Carlson, 1959), were called pseudoalleles. Genes containing pseudoalleles became known as complex loci. Doubts were expressed regarding the functional test for allelism, and as a result the phenotypic difference observed between the *cis* and *trans* compounds was called a position effect of pseudoallelic type (Lewis, 1945).

Pontecorvo (1950) and Lewis (1951) regarded pseudoalleles as individual subgenes closely linked with each other functionally. It was postulated that for them to function successfully pseudoalleles must be next to each other on the chromosome (the *cis* configuration). If they are on different chromosomes (the *trans* configuration), gene products participating in the general cycle of biochemical reactions cannot interact with each other, the cycle is broken, and complementation does not take place.

For an examination of the present state of the pseudoallelism problem, see Lewis (1967) and Grace (1970).

2.4. Intragenic Recombination

The low resolving power of genetic analysis and the complexity of the characters available for study prevented a solution to the problem of pseudoallelism from being found by work on *Drosophila*. By chance, at this time, a number of microbiological systems in which, by contrast with higher organisms, the biochemical characters were precisely controllable and millions or billions of progeny could be obtained easily from one pair of individuals, were being intensively studied.

Bonner (1951) showed that not only the genes determining the complex morphological character in *Drosophila,* but also the genes controlling specific biochemical reactions in *Neurospora crassa* are complex in nature. He distinguished a series of mutations, each of which blocked the conversion of 3-hydroxyanthranilic acid into nicotinic acid, as a result of which the strains carrying these mutations accumulated quinolinic acid. By crossing these strains with each other Bonner could easily (because he used a selective medium on which only wild-type recombinants could develop) observe recombination between these mutations, which occured in a frequency of $2 \times 10^{-4} - 5 \times 10^{-4}$.

Similar results were obtained by Roper (1950) in the analysis of biotin (*bi*) mutants in *Aspergillus nidulans*. He even compiled a map of the mutual arrangement of the three pseudoalleles of the *bi* locus,* showing its orientation with respect to the whole linkage group by means of flank markers. The functional test for allelism showed that these mutations are not complementary (Roper, 1953).

On the basis of these results Pontecorvo (1952) put forward a remarkable suggestion. He rejected the dogmatic view of indivisibility of the gene and postulated that pseudoalleles are true alleles responsible for various mutation injuries in different sites of the long but functionally whole gene.

*The term "locus" in modern genetics no loger has its classical meaning. It now implies not only the site of the gene on the chromosome, but also the region where phenotypically similar mutants between which the functional relationships are unknown, not understood, or of no significance, are mapped. A synonym of this term is "series" (3.4).

In brief, he postulated that recombination between pseudoalleles is in fact intragenic (interallelic) recombination. This idea was brilliantly confirmed by the whole subsequent course of development of molecular genetics. The term pseudoalleles had to be rejected because it was based on the concept of indivisibility of the gene, which has been found to be mistaken. Roman (1956) introduced the term "heteroalleles" for alleles capable of recombination, and the term "homoalleles" for those which cannot recombine with each other.

2.5. The Structure of the Gene

According to Pontecorvo's hypothesis, the gene as a functional unit does not coincide with the units of recombination or mutation. The fact that recombination takes place between alleles means that the gene has a complex linear structure. What is the nature of this structure? What is the limit of divisibility of the gene? To what do the recombination and mutation units correspond physically? Do they coincide? In order to answer these questions it was first necessary to draw a detailed map of a gene. However, it could legitimately be asked whether recombination frequencies can be used for the construction of such a map as had been done for mapping chromosomes. Since this problem could not be solved *a priori*, there was only one way out of the difficulty: to adopt as a working hypothesis that during crossing of allelic mutants, wild-type recombinants are formed by crossing-over. If this is so, the laws established by Morgan's school for recombination of genes would be completely applicable to intragenic recombination also. If the gene, like the chromosome, is a linear structure, drawing a map of the gene would be a very simple matter. In fact, the linear arrangement of genes on the chromosome was deduced (Chapter 1) mainly from the additivity of the *rf* values over short distances. The shorter the distances, the better the additivity, for the probability of double exchanges is reduced. Consequently, additivity inside the gene must be at a very high level. The possibility of compiling gene maps on the basis of this assumption was demonstrated by the very first investigations of pseudoallelism (2.3).

The task of differentiating the three functional aspects of the gene was finally solved by Benzer (1955, 1957) on bacteriophage T4 of *Escherichia coli*. The use of a bacteriophage system was determined by the need for developing methods of finding rare recombination events, for recombination between alleles of the same type, as we have already seen, is extremely infrequent.

Benzer found that *rII* mutants of phage T4 growing in cells of *E. coli* B do not form a mature phage progeny in the cells of *E. coli* K12 (λ), although they infect and kill the bacteria of this strain. The "lawn" of bacteria of this strain on a Petri dish behaves as a selective medium on which only wild-type phages (r^+) and not *rII* mutants will grow. Making use of this property of the *rII* mutants Benzer developed a highly sensitive test by which he could easily observe rare genetic events taking place within this region of the phage genome: recombinations and back-mutations. The sensitivity of this method in practice is limited only by the frequency of reversions of *rII* mutants to the wild type. The frequency of the reversions is determined by seeding known large numbers (up to 10^9) *rII* phage particles on a lawn of *E. coli* K12 cells. This seeding also serves as a control recording the "noise" level in recombination experiments. As has already been emphasized, in order to solve the problem it was necessary to compile the most detailed possible map of a gene, i.e., to fill it to the limit with mutations. For this purpose Benzer isolated more than 2400 spontaneous and induced mutations in the *rII* region. On the basis of a test for allelism described by Benzer as the *cis–trans* test these mutants were distributed among two functionally independent regions A and B, which Benzer called "cistrons." The term cistron at one period had a firm foothold in the genetic literature, but more recently, because of the discovery of interallelic complementation (Fincham, 1968). and with the consequent inadequacy of the functional test for allelism alone, there is a tendency to go back to the classical term "gene" (Hayes, 1968).

The functional test for allelism is based on mixed infection of *E. coli* K12 by two *rII* mutants of independent origin. If the multiplicity of infection (the number of phage particles per bacterial cell) is large enough, particles of both types will enter most cells. This creates a situation similar to the compound in higher organisms and it allows the presence or absence of complementation to be noted. If the injured sites belong to different functional units (genes), the phages develop normally. If the mutations are allelic, the phage does not develop, lysis of the culture does not take place, and the number of sterile plaques remains at the control level.

The usual two-point cross method was used (the term two-factor or, still more, two-gene cross is meaningless when applied to intragenic recombination). The crosses were carried out on *E. coli* B and the total number of phage particles ($r^+ + rII$) in the suspension after lysis was determined by seeding the suspension in an appropriate dilution on a lawn of *E. coli* B and the number of wild-type (r^+) recombinants was determined separately by seeding the same suspension undiluted on a lawn of *E. coli* K12; *rf* was determined from the ratio between the titer of sterile plaques on *E. coli* K12 to the titer on *E. coli* B.

Since Benzer began with a crossing-over model and no reciprocal recombination products (double mutants) were found on strain K12, to obtain the true recombination frequency this ratio was doubled.* To compile a complete genetic map all possible pairs of mutants had to be crossed, a truly

---

*As we shall see later (Chapter 3) this is incorrect. The value of *rf* in interallelic crosses is now generally expressed in frequencies of appearance of wild-type recombinants or tetrads containing recombinant spores (w); in the latter case $rf = 0.25w$.

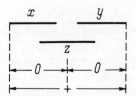

Fig. 18. The method of overlapping deletions (from Benzer, 1957). If recombinants of wild type (O) are not formed during crossing of mutant z with mutants x and y, the deletion z occupies a central position (provided that x and y recombine with each other).

impossible task. In fact, the number of possible crosses in pairs is given by $n(n-1)/2$, which is more than 3 million if there are 2500 mutants.

To facilitate the work Benzer developed a method of preliminary (qualitative) mapping by the use of deletions (Fig. 18). The results obtained by paired crossing of mutants of deletion type enable the only possible mutual arrangement of the deletions to be determined. If a set of deletions of different lengths, covering the whole gene locus, is available, the task of

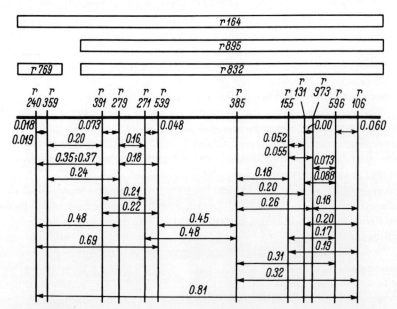

Fig. 19. Map of segment r164 of gene rIIA of phage T4 (from Benzer, 1957). Numbers above the map are serial numbers of alleles; numbers below the map are values of $rf \times 10^2$, determined in corresponding two-site crosses (alleles participating in the cross are joined by lines terminating in arrows).

§2.5] STRUCTURE OF THE GENE

mapping a newly isolated point mutation amounts initially simply to the discovery of the position of the deletion which covers that particular point: the mutation to be studied is crossed with a deletion covering a considerable part of the gene. If the mutation is localized in this region, no wild-type recombinants will appear. Crossing is then carried out with deletions of successively decreasing size, located in this region of the gene. By this method, preliminary grouping of point mutations can be achieved. Later crosses between all possible pairs of point mutants are carried out only within each group. The arrangement of the sites on the map within these groups was determined from the observed recombination frequency in the paired crosses.

A map of one segment of the *rIIA* gene overlapped by deletion *r164* is shown in Fig. 19. Despite the low level of additivity (for further details, see 2.7), the linear arrangement of the sites is demonstrated sufficiently clearly because linearity can be established on the basis of a criterion not depending on additivity (Tessman, 1965). Suppose we wish to determine the order of arrangement of the sites on the basis of two-site crosses. Let us determine rf_{ab}, rf_{bc}, and rf_{ac} in corresponding crosses between pairs of mutants. The postulate of linearity states: a mutation not participating in the cross in which rf_{max} was recorded occupies a midposition on the map. If in our example $rf_{ab} > rf_{bc}$ and $rf_{ab} > rf_{ac}$, i.e., rf_{ab} is rf_{max}, site *c* must occupy the central position. Application of this criterion consecutively to the group of mutations *a, b, c, ..., x, y, z* enables the mutations to be arranged in linear order without the need for observance of the law of additivity.

More strict proof of linearity is obtained by the three-point cross method, the traditional genetic test for linearity (Fig. 20). Chase and Doermann (1958), who used this method later for a detailed mapping of a limited number of mutations, confirmed the linear character of site arrangement in the *rII* region. The arrangement of the markers is determined absolutely unequivocally by this method, but a disadvantage is that double mutants must first be obtained and this makes the investigation much more difficult.

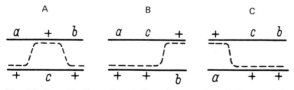

Fig. 20. Localization of mutations by a series of three-point crosses. If the order of the sites is *acb*, the frequency of appearance of wild-type recombinants in cross A is much lower than in crosses B or C. A disadvantage of this method is that double mutants (*ab, ac,* and *cb*) must be obtained beforehand.

It was thus shown that the sites in segments defined by deletions are arranged in linear order. By mapping mutations at the junctions between deletions or in regions of overlapping, all segments of the gene could be linked together. The absence of contradictions when sites were mapped by the deletion method and the coincidence between the results obtained by this method and those of mapping by *rf* values completely ruled out the possibility of branching of the map. The linear structure of the gene has now been proved not only for microorganisms, but also for *Drosophila* (Welshons and Von Halle, 1962; Chovnick et al., 1970), *Zea mays* (Nelson, 1962), and other objects.

2.6. The Recon

Benzer thus showed that the gene can be mapped and that it is linear down to the lowest level. It was very tempting to convert genetic distances into physical distances by relating size on the genetic map to length of the phage DNA. Benzer attempted to do this. For the unit of mutation Benzer proposed the name "muton," and for the unit of recombination "recon." The muton is the smallest element of genetic material through a change of which a mutation can arise. The recon is the smallest element which is exchanged during recombination but does not itself divide (the "atom" of recombination). What is the size of these units? What is the limit of divisibility of the gene? By simple calculation the order of their magnitude can be determined. Having obtained a value of 800 map units for the total length of the genetic map of the phage and having compared it with the number of nucleotide pairs in the genetically active part of the DNA molecule of phage T4 (at that time it was considered that only 40% of phage DNA, or 8×10^4 nucleotide pairs, carries genetic information), Benzer found that 10^{-2} map unit corresponds to 1 base pair.* Since the recombination frequency least different from zero observed in his experiments was 2×10^{-4} (although the sensitivity of his system was such that recombination events taking place at a frequency at least one order of magnitude lower could be determined), Benzer concluded that the recon cannot be larger than 2 base pairs in size. Correspondingly, the size of the muton cannot be larger than this.

Absolute proof that the recon defined by Benzer is a nucleotide pair was obtained by Yanofsky et al. (1964) by biochemical and genetic analysis of the tryptophan synthetase system in *E. coli*. Yanofsky and his collaborators studied the effect of mutations in the A gene on changes in the primary structure of the protein A subunit of the enzyme tryptophan synthetase. The mutant proteins, which differed from the normal in only one

---

*A calculation based on more up-to-date results gives the same value: the whole chromosome of the phage (2×10^5 nucleotide pairs) is genetically active and its length is 2000 map units.

amino acid, were identified chromatographically by the "fingerprint" method. Attention was concentrated on those groups of mutants in which amino acid substitutions were found to take place at the same link of the polypeptide chain. These mutations were first mapped (by the aid of transduction) in the same site of the *A* gene, but a more refined analysis revealed recombination between them at an extremely low frequency (10^{-5}). In one mutant (*A46*) the glycine of peptide I, found in the normal protein, was replaced by glutamic acid, while in another (*A23*) the same glycine residue was replaced by arginine. This meant that the mutation in both cases took place in the same coding triplet (codon). Analysis of the A protein isolated from wild-type recombinants showed that it does not differ from normal; this meant that the normal sequence of nucleotides had been restored at this site, coding glycine, by recombination (Fig. 21A). An example similar to that described above was found by the study of another pair of mutants (*A58* and *A78*) in which the same glycine residue in peptide II was replaced by aspartic acid and cysteine respectively. Recombination was also observed between these mutants with a frequency of 1×10^{-5}, with restoration of the

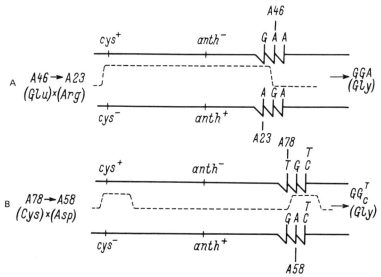

Fig. 21. Intracodon recombination (after Yanofsky et al., 1964). A) Transduction of *A46* allele into bacterial cell carrying mutation *A23*. B) Cross *A78* → *A58*. Broken lines denote genotype of selected recombinants, *cys* and *anth* denote genes controlling synthesis of cysteine and anthranilic acid. Glu, Arg, Gly, Cys, Asp — corresponding amino acid residues (glutamine, arginine, glycine, cysteine, and asparagine) in the polypeptide. G, A, T, C — guanine, adenine, thymine, and cytosine.

normal amino acid sequence in A protein and with the formation of wild-type recombinants (Fig. 21B).

The results obtained by Yanofsky and his collaborators proved that recombination can take place inside the codon and, consequently, that the recombination unit is one nucleotide pair. However, the frequency of recombination within the codon was not the same in every case. In the examples discussed it was of the order of 1×10^{-5}, while in other cases it was rather higher, up to 3×10^{-5}. Moreover, the attempt to estimate the value of the unit of genetic recombination from Yanofsky's results by a calculation similar to that used by Benzer gives paradoxical results. The polypeptide chain of A protein is known to contain 280 amino acid residues, which corresponds to 840 nucleotide pairs in the A gene (assuming a triplet code). The total length of the A gene is 4 map units and the minimal recombination frequency is 1×10^{-5}. This means that in the A gene there must be about 4000 nucleotide pairs ($4 \times 10^{-2}/10^{-5}$), i.e., almost five times more than are required. This was the first indication that rf does not correspond unequivocally to physical distance at the intragenic level. Meanwhile, using the three-point cross method, Yanofsky et al. also demonstrated that the arrangement of the sites within the codon may also be linear with respect to the chromosome as a whole. Their experiments finally confirmed Benzer's conclusion that the physical limit of divisibility of the gene is one nucleotide. In fact, since the sugar and phosphoric acid residues in DNA have evidently no coding functions, this is as it should be.

If rf is a simple function of physical distance, the value $rf = 1 \times 10^{-4}$ is the minimum for phage T4. As Benzer puts it, it is the "absolute zero" for rf: "If not even 0.01% of recombinations is obtained by crossing two single mutants of phage T4 it can be concluded that the distance between the sites of these two mutations is less than 1 nucleotide pair" (Benzer, 1957, p. 87). Benzer's conclusion was categorical enough but it was wrong.

The study of recombination at "hot spots" of phage T4 led to a surprising discovery. Before a description of these experiments is given it will be useful to examine a special feature of gene maps, namely, the character of their topography. If mutations within region bounded by the ends of deletions are crossed, many point (reverted) mutations do not recombine. Benzer naturally located these mutations at the same site and regarded them as repeated mutations of the same muton. Altogether more than 300 individual sites were identified in the rII region inside which about 2500 mutations were localized; there are thus 8 repeated mutations on the average per site. Nevertheless, the site distribution of the mutations found differs sharply from normal. In most cases the site is represented by only one mutation, but on the other hand there are some sites in which several hundred mutations are localized: these sites were called "hot spots" (for example, about 300 mutations are localized in the hot spot in segment $A6$ of gene A and more than 500 in segment *B4* of gene *B*). It is interesting to note that the hot spot spectrum was found to differ for classes of mutations induced by different mutagens. The same unequal distribution of the frequency of mutations by sites and hot spots along the segments of the gene has been found in many other genes both of phages and of fungi and bacteria (Holliday, 1964b; Sukhodolets et al., 1965). The structure of one hot spot in the lysozyme gene of phage T4 has been worked out: it is a sequence of six identical bases (Okada et al., 1972).

Tessman (1965) developed an ultrasensitive method of finding rare recombination events. This method differs from Benzer's standard method in a very simple modification. Cells of *E. coli* B infected by test mutants of phages are not allowed to undergo lysis but are seeded directly onto a lawn of *E. coli* K12. This modification reduces the "noise" level: instead of large clones of revertants and small clones of recombinants, standard sterile plaques are obtained equally. Tessman's method, by facilitating discrimination of recombinants and the mutational background, increases the resolving power of the genetic analysis by more than two orders of magnitude, i.e., it allows the detection of recombination events taking place with a frequency of $10^{-7}-10^{-8}$ (compared with $10^{-5}-10^{-6}$ with Benzer's method). Of 648 mutations induced by nitrous acid in the *rII* region 157 were localized by the standard method in one site (hot spot). It was later found that the mutations of this group can be divided into three subgroups on the basis of "phenotype." To characterize the phenotype two criteria were used: the frequency of reversions to normal and suppressibility, i.e., ability to reproduce on certain mutants of strain *E.coli* K12 possessing "suppressor genes." Attempts to cross phenotypically different mutants from the same hot spot have proved successful. A hot spot has been "resolved" into three sites by Tessman's method (Fig. 22).

All mutations belonging to the same phenotypic group and localized by Tessman at the same site in the hot spot behaved identically. Phenotypically homogeneous sites could not be further resolved by this method. Moreover, mutations induced by hydroxylamine in these three sites recombined with the same frequencies. What conclusions must be drawn from these observations? First, it is evident that the sites in the hot spot correspond to three adjacent nucleotides. Second, since the *rf* values of neighboring nucleotides may differ by an order of magnitude and additivity in general is absent at these distances, it is impossible to establish a precise relationship between *rf*

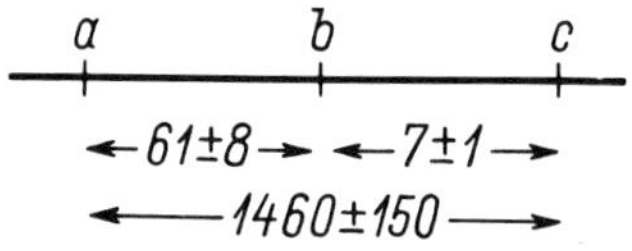

Fig. 22. Map of "hot spot" in gene *rIIB* of phage T4 (after Tessman, 1965). Numbers represent $rf \times 10^8$.

$a \quad b \quad c$
$\leftarrow 61\pm8 \rightarrow \leftarrow 7\pm1 \rightarrow$
$\leftarrow\!\!\!-\!\!\!-\!\!\!- 1460\pm150 -\!\!\!-\!\!\!-\!\!\!\rightarrow$

and the recon. The important conclusion can thus be drawn that in maps of a high level of accuracy *rf* is not a measure of physical distance. In fact, *rf* values from 1×10^{-5} to 1×10^{-1} have been found in the *rII* region. If *rf* corresponded directly to physical distance (1 recon = 1 nucleotide pair = 10^{-5} *rf*) the *rII* region alone of phage T4 would have to contain 1×10^7 base pairs, although there are only 2×10^5 base pairs in the phage genome. Facts of this type have continued to accumulate. For example, in full agreement with the findings of Tessman (1965), Ronen and Salts (1971) found that *rf* in the 12 internucleotide intervals of phage T4 varies from 6.2×10^{-7} to 5.4×10^{-5}. In a study of intracodon recombination in the *lac* region of *E. coli* (Zipser, 1967) *rf* values between adjacent nucleotides were 10 times less than those expected theoretically on the basis of the assumption of invariance of the physical scale of the gene map. These facts brings us straight to the problem of additivity.

2.7. Additivity within the Gene

In the first stages of the study of intragenic recombination it was hoped that the basic law of genetic recombination would apply very precisely when the gene was mapped. However, the very first experiments on intragene mapping showed that, despite the linear arrangement of sites in the gene, additivity of the distances between them is not observed. It seemed at first that these deviations from the recombination law could be due to various technical causes (Chase and Doermann, 1958). In fact, since the *rf* values between different pairs of alleles are determined in different experiments, some degree of error must inevitably arise. Besides fluctuations in *rf* due to uncontrollable changes in the physiological state of the microorganisms, which could be eliminated to some extent by standardization (although this could be done only in merozygotic systems), the technical conditions under which the experiment is performed are of tremendous importance. For example, the multiplicity of infection has a considerable effect on recombination frequency in phages (Mosig, 1962). Another factor with a marked effect on *rf* in interallelic crosses is the genetic constitution of the cross strains (in eukaryotes). For example, in experiments on *N. crassa* Murray (1963) found a very high level of additivity on crossing of the alleles in the *me-2* (methionine synthesis) gene. Nevertheless Murray remarks that these results may be purely accidental, for *rf* between alleles is very highly dependent on the viability of the spores and also on differences in the genetic constitution of the strains. The use of isogenic strains leads to a decrease in *rf* between alleles belonging to the α and γ groups from 2.5×10^{-3} to 2.5×10^{-4}. Reciprocal crosses using genetically heterogenous strains often give results which differ by several times (Murray, 1963, Table 4, cross 3). The significance of these observations has recently been made clear by the work of Catcheside's group (3.8).

One of the most important sources of error when mapping the gene is allele-specific recombination (2.12). In practically every paper which has been written on mapping the gene there is mention of mutations with abnormal recombination behavior and which cannot be localized by the two-point cross method (Pateman, 1958; Edgar et al., 1962; Helinski and Yanofsky, 1962; Lissouba et al., 1962; Touré and Picard, 1972). It is not surprising, therefore, that the results obtained by different workers on the same system or, still more, on different systems are so contradictory and heterogeneous. Frequently when the alleles of one gene in the same system are mapped, both strict additivity and a deviation from it in both directions may be found (Smith, 1965; Fincham, 1967). In some cases the level of additivity was found to be very high (Ishikawa, 1962), in others very low (Lawrence, 1956; Suyama et al., 1959), and in some cases it was actually impossible to determine the order of the sites within the locus (Smith, 1961; Stadler and Kariya, 1969).

Virtually all investigators have had to contend with difficulties when mapping genes entirely on the basis of two-point crosses, and it is therefore essential to use three-point crosses or the flank marker method in order to determine the precise location of sites in the gene. If all distorting factors have been eliminated (abnormal mutants discarded, genetically homogeneous strains crossed, technical conditions standardized, adequate statistical material, order of the sites confirmed by additional methods), clear deviations from additivity are nevertheless observed, and in a completely unexpected direction. Let us go back to the experiments on intracodon recombination described earlier (2.6). The marked difference in rf between neighboring nucleotides leads to a deviation from additivity which is called "widening of the map": the sum of the intermediate rf values is less than rf between the sites at the borders of the interval chosen ($rf_{ab} + rf_{cb} < rf_{ac}$). This effect is characteristic not only of internucleotide recombinations, but also of any other recombinations within the gene (Fig. 23). It was Holliday (1964b) who first drew attention to the existence of widening of the map. The widening effect is not universal and it is manifested only as a tendency, especially if the results are examined collectively. By more detailed analysis we find three levels of recombination within the gene, depending on the number of nucleotides concerned: recombinations between fewer than ten nucleotides are characterized by considerable widening of the map; between 10 and 100 nucleotides additivity of rf is reasonably good, but with recombinations of more than 100 nucleotides widening again takes place. Widening of the map evidently occurs if deletions are mapped. (If in a group of mutations a, b, and c the midposition is occupied by deletion b which is 1 map unit long and if $rf_{ab} = rf_{bc} = 0.2\%, rf_{ac} = 1.4\%$, i.e., it is several times greater than the expected value of 0.4%.) However, in most cases point mutations were studied.

The pattern observed in bacteriophage T4 differs from that in fungi. Benzer found "narrowing" of the map in the rII region: $rf_{ab} + rf_{bc} > rf_{ac}$

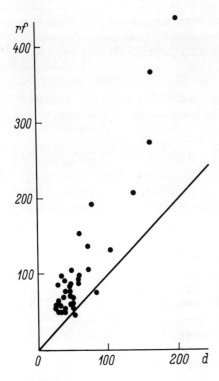

Fig. 23. Relationship between *rf* and *d* in the ad_6 gene of *Schizosaccharomyces pombe* (from Holliday, 1964b). Coordinates of the experimental points determined by a standard method (Fig. 4). Continuous line shows curve described by Sturtevant's function; $rf \times 10^6$; $d \times 10^6$.

(Fig. 19). The deviations from additivity are so great that the length of region *rII*, determined by summation of the short intermediate intervals, is several times greater than the recombination frequency of the terminal sites. It was subsequently shown (Stahl et al., 1964; Fisher and Bernstein, 1965) that deviations of this nature from additivity are characteristic of all regions of the phage genome studied (Fig. 24). What is the reason for the absence of widening of the map for T4? This effect may perhaps be masked by low negative interference due to the population character of phage genetics (1.18). Another masking factor could be the phage heterozygotes (5.6). Doermann and Parma (1967) showed that if *rf* is determined under selective conditions the values obtained are too high, because heterozygotes are taken for recombinants. Since the frequency of their appearance is inversely proportional to distance, the closer the corresponding alleles are arranged the greater the error in determination of *rf* (for another explanation, see page 193).

Interesting results have been obtained by the study of intragenic mitotic recombination in yeast induced by x-rays (R. E. Esposito, 1968). In this case widening of the map was replaced by narrowing at large intragenic distances. In the case of γ-induced mitotic recombination in the *ilv-1* gene in yeast no widening of the map has in general been found (Thuriaux et al.,

1971). If this effect is subsequently confirmed and more convincing statistical evidence is obtained it could indicate differences between the mechanisms of spontaneous and induced recombination.

Deviations from additivity in interallelic crosses are thus found for both relatively long and very short distances. Distances inside hot points are absolutely nonadditive (2.6). Furthermore, the sites in them cannot be arranged in linear order relative to the gene map, even on the basis of three-point crosses (Tessman, 1965). Sites can be arranged within the codon only if the third marker is some distance away in the same gene (Zipser, 1967) or in the neighboring gene (Fig. 21).

2.8. The Flank Marker Method

Because of the difficulties which often arise in mapping sites within the gene solely on the basis of determining *rf* in two-point crosses, qualitative mapping methods have had to be used. One such mapping method is that based on the use of flank markers. However, this method can only be used with systems in which closely linked markers are present on both sides of the gene within which recombination is to be studied. They are usually called outside or flank markers. The use of the flank marker method for intragenic

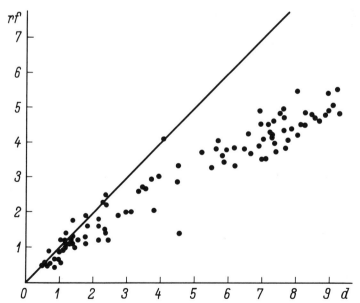

Fig. 24. Relationship between *rf* and *d* in the *rII* region of phage T4 (from Fisher and Bernstein, 1965). Coordinates of the points obtained by different workers are determined by the standard method (Fig. 4).

mapping is based on the following assumption: if the linear gene map is a direct continuation of the chromosome map, crossing-over within the gene must lead to recombination of the flank markers. If such markers are present in the immediate vicinity of the gene boundaries, so that the possibility of additional exchanges affecting the flanks is reduced, the order of the mutant sites within the gene can be determined relative to them. On the basis of the classical scheme of crossing-over the character of distribution of flank markers among selected recombinants can be predicted if the distances between the flanks and the investigated gene are known (i.e., the distances $M-a$ and $N-b$, where M and N are the flank markers and a^+ and b^+ the selected alleles).

The scheme of crossing between allelic auxotrophic mutants (a and b) with the participation of flank markers (M and N) and the possible variations of their distribution in selected a^+b^+ recombinants is shown in Fig. 25. Let it be assumed that $rf_{Ma} = 0.05$ and $rf_{Nb} = 0.01$. If the order of arrangement of the markers is as in A, most prototrophs must possess the mN recombinant combination of flank markers, for one crossover in region 2 is essential for their formation. The MN combination of flank markers must be possessed by 5% of wild-type recombinants, for an additional exchange in region 1 is required for the formation of this configuration. Correspondingly, the MN combination of flank markers must be found among a^+b^+ in 1% of cases (an additional exchange in region 3). Finally, a^+b^+ recombinants with the Mn configuration of flank markers must be found in only 0.05% of cases, for two additional exchanges are required in regions 1 and 3 for its formation.

If most wild-type recombinants obtained from an $a \times b$ cross (Fig. 25) possess the mN flank configuration, mutant sites within the gene must be arranged in the order A. It will easily be seen that if the markers are arranged in the opposite order (B) most prototrophs will possess the Mn configuration

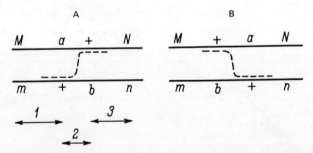

Fig. 25. The flank marker method. In the cross $MaN \times mbn$, where a and b are closely linked markers (usually alleles of the same gene, but they may also be nonallelic) a^+b^+ recombinants are selected. Markers $M(m)$ and $N(n)$ are unselected. A) Order of markers $MabN$; B) order of markers $MbaN$.

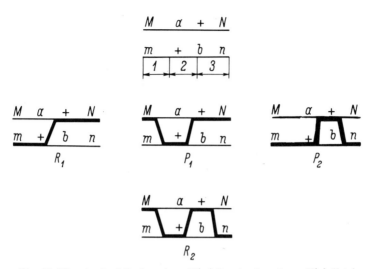

Fig. 26. The standard flank system. $M(m)$ Proximal markers, $N(n)$ distal markers. Intervals: 1) proximal, 3) distal. Thick line denotes genetic structure of recombinants.

of flank markers.* In either case marked differences must be observed both in the relative frequencies of appearance of the two recombinant (R) flank classes and of the R and P classes: there must always be fewer P than R. Crosses in which recombinants were selected with respect to two closely linked nonallelic mutations (Stadler, 1956; Calef, 1957; Giles et al., 1957) fit into this scheme completely. A high frequency of one of the R classes was observed, but the other R class was virtually completely absent, and the frequency of appearance of the P classes was low or very low. However, in the very first experiments in which the distribution of flank markers was studied in recombinants obtained in interallelic crosses, marked deviations from the theoretically expected frequencies were found. By crossing alleles of the *inos* gene (inositol), Giles (1951) found for the first time that P classes appear at a frequency commensurate with that of the appearance of R classes. Pritchard (1955) made frequent crosses of different alleles of the ad_8 gene (adenine synthesis) of *Aspergillus nidulans*. The parent strains were marked with the y (yellow conidia) and bi (biotin) genes, located on either side of ad_8 at distances of 0.2 and 5.7 map units respectively ($y\,ad_x \times ad_y\,bi$). Instead of the expected 6%, the frequency of P combinations of flanks in the recombinants selected on medium without adenine reached 40% in

---

*Since we shall use the flank systems in future as the source of other information than site order, this is a convenient point to introduce the standard flank system (Fig. 26), in which P_1 denotes the flank configuration of the parent with the proximal allele a, P_2 the configuration of the parent with the distal (b) allele, R_1 the flank configuration arising by crossing-over in the region between the alleles, and R_2 the configuration arising as a result of additional exchanges on both flanks.

some crosses (in the *methA* gene of *A. nidulans* the frequency of P_1 reaches 75%; Putrament et al., 1971). Although Pritchard found that one R class was predominant over the other, in no case was the difference between them as large as ought to be expected. Furthermore, the frequencies of classes R_1 and R_2 in *N. crassa* were in many cases almost equal (Serres, 1956; Freese, 1957; Bausum and Wagner, 1965). Equality of the R_1 and R_2 classes naturally gave no indication of the order of arrangement of the sites.* Fortunately this paradoxical situation is not always found. Asymmetry in the distribution of the R classes is observed in most crosses, and on the basis of this asymmetry the sites within the gene can be arranged in linear order to coincide with the order determined by two-point or three-point crosses. Although there are some difficulties, the flank marker method does give good results, even in cases in which the allele behaves aberrantly and cannot be localized on the basis of the *rf* value determined in two-point crosses.

How can the anomalies in the distribution of the flank markers be explained; Why do P and R_2 classes appear in excessive numbers? The situation would appear to be formally as if exchange between alleles increases the probability of appearance of additional exchanges at the flanks (Giles, 1951) and consequently the observed character of distribution of the flank markers can be described in terms of interference. Pritchard (1955, 1960a,b) called this phenomenon local negative interference. Negative interference can also be observed in three-point crosses.

2.9. High Negative Interference

Chase and Doermann (1958) undertook a series of three-point crosses (having first obtained double and triple mutants from Benzer's strains) in order to obtain stricter proof of the linear arrangement of sites within the *A* and *B* genes in the *rII* region of phage T4 and hit upon a completely unexpected fact. They crossed the double mutant *r168r147* with the single

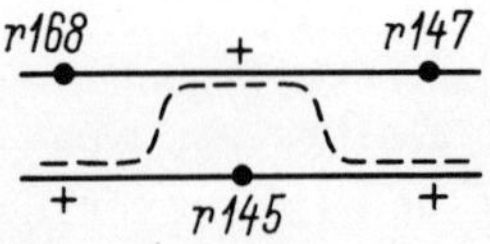

Fig. 27. Three-point cross involving *rII* mutants of phage T4.

---

*In these cases the order can still be established on the basis of two additional criteria (Jessop and Catcheside, 1965): 1) if, in a^+b^+ recombinants, only proximal flank alleles (*M* and *m*) are taken into account, if $m > M$ the order of the markers is *Mab*, while if $M > m$ the order of the markers is *Mba*; 2) if distal alleles are considered if $N > n$ the order of the markers is *abN*, and if $n > N$ the order is *baN*; if the criteria give consistent results the reliability of determination of the order can be guaranteed.

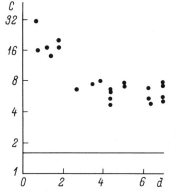

Fig. 28. Coincidence factor (C) as a function of distance between terminal sites (d, in map units) (from Chase and Doermann, 1958). Line in lower part of graph shows level of low negative interference (see 1.18).

mutant *r145* (Fig. 27). The approximate localization of these sites was established by two-point crosses. As the diagram shows, wild-type (r^+) recombinants can be formed only by double crossing-over; consequently, the frequency of their appearance in the progeny must be the product of the frequencies of appearance of wild-type recombinants in two-point crosses ($rf_{168-145} \times rf_{145-147}$). These frequencies were known: 4.1×10^{-3} and 1.4×10^{-3}. The probability of appearance of the double recombinant was therefore 5.7×10^{-6}. However, the frequency of appearance of r^+ was found to be 1.8×10^{-4}, which is 30 times higher than the expected frequency ($C = 30$). The results obtained by these workers, illustrated in Fig. 28, show that C depends on the distance between the terminal sites involved in the three-factor cross. If both terminal sites are located in the same gene, C varies around 30. In crosses in which terminal sites in neighboring genes (A and B) were used, the value of C was much lower, but as before it was higher than $C = 1.6$, which is characteristic of intergenic crosses in phage T4 (1.18). Chase and Doermann called this phenomenon high negative interference, in order to distinguish it from the low negative interference ($C = 1.6$) exhibited in phage T4 during intergenic crosses.

Pritchard also found that the degree of negative interference is highly dependent on the closeness together of the flank markers: the more closely they are linked the higher the degree of negative interference. Pritchard therefore called it local negative interference also. The existence of this type of interference has been demonstrated in all systems so far studied (Sturtevant, 1951; Kaiser, 1955; Hexter, 1963; Amati and Meselson, 1965; Jacob and Wollman, 1961, p. 295; Norkin, 1970; Salamini and Lorenzoni, 1970) except transformation systems (Gray and Ephrussi-Taylor, 1967; Morse and Lerman, 1969). There is likewise no strict evidence that it exists in the case of transduction (Gross and Englesberg, 1959). Its presence has been demonstrated only recently by the use of a flank system (Crawford and Preiss, 1972).

2.10. The Effective Pairing Hypothesis

With the discovery of the phenomenon called high negative interference it became evident that the theory of genetic recombination no longer applied in its orthodox form. The linear scales used by geneticists when compiling their genetic maps are noninvariant. They contract for the measurement of long distances and they stretch for the measurement of short distances. Attempts to explain this phenomenon have been made in various ways.

Pritchard (1955) abandoned the classical position and postulated a single mechanism of genetic recombination, on the basis of which he developed a simple modification of the crossing-over hypothesis which had been suggested initially by Rothfels (1952); the basic assumption here was that recombination between genes is merely the result of events taking place on very short segments of chromosomes, known as regions of effective pairing. The classical theory regarded synapsis of chromosomes in meiosis as effective and continuous contact along their whole length (Fig. 29). According to the effective pairing hypothesis, synapsis between homologues is not complete but interrupted. Since effective pairing of two homologous chromosomes is essential for crossing-over, crossing-over is possible only within the segments where contact is made between the chromosomes. These segments were called "switching regions" (Chase and Doermann, 1958) or regions of effective pairing (Pritchard, 1955).

According to this hypothesis the frequency of recombination between genes is not a true expression of the frequency of crossing-over. The latter is much higher in regions of effective pairing and it tends toward the values determined in three-point crosses or on recombination of flank markers (2-3 exchanges per region of effective pairing – see below). This means that rf between genes reflects principally the probability of formation of regions of effective pairing between them. At the same time, recombination of alleles requires that they both fall within the region of effective pairing. Consequently, rf between alleles is the product of two probabilities: the probability of formation of a region of effective pairing in a given locus of the chromosome (p_O) and the true frequency of crossing-over between alleles in the region of effective pairing (p_x), or $rf = p_O p_x$.

Fig. 29. Types of pairing of homologous chromosomes: A) complete pairing (the classical scheme); B) interrupted pairing (Pritchard's scheme). The chromosomes in both schemes are shown on the same scale. Crosses denote crossings-over. Their frequency per unit physical length of chromosome is much higher in Pritchard's scheme.

§2.10] EFFECTIVE PAIRING HYPOTHESIS

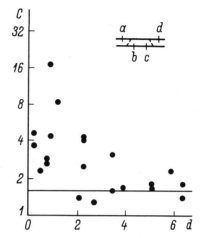

Fig. 30. Determination of size of region of effective pairing in a series of four-point crosses (from Chase and Doermann, 1958). Scheme of crossing shown in top right-hand corner of Figure. Abscissa, distance between markers b and c. Remainder of legend as in Fig. 28.

For simplicity let p_x be equal for two pairs of alleles ab and bc. In that case $rf_{ab} = p_o p_x$ and $rf_{bc} = p_o p_x$. In a three-point cross $ac \times b$ the frequency of appearance of wild-type recombinants must be given by the formula $p_o(p_x)^2$, in which p_o occurs only once. However, we calculated it by the equation $rf_{ab} \times rf_{bc} = (p_o)^2 (p_x)^2$. High negative interference thus arose

$$C = \frac{p_o (p_x)^2}{(p_o)^2 (p_x)^2} = \frac{1}{p_o},$$

since $p_o < 1, C > 1$. Hence, if we compare rf in two-point and three-point crosses we are dealing with strikingly standardized values. By restandardizing two-point and three-point rf values with respect to the probability of formation of a region of effective pairing we can eliminate high negative interference.

Two important parameters which can be determined experimentally are introduced into Pritchard's hypothesis: the size of the region of effective pairing and the probability of crossing-over per unit length within this region (the true frequency of crossing-over). The length of the region of effective pairing can be measured in a series of four-point crosses (Chase and Doermann, 1958) (Fig. 30). In crosses of this type the value of C must decrease with an increase in the interval $b-c$. Disappearance of high negative interference must take place if the interval $b-c$ is larger than the region of effective pairing. As the results obtained by Chase and Doermann show (Fig. 30) the degree of high negative interference diminishes rapidly with an increase in the distance between the sites $b-c$ and falls to the normal value of low negative interference if the distance exceeds 4%. This corresponds approximately to the mean size of the gene of phage T4. The size of this region in phage λ is similar (1.5×10^3 nucleotide pairs) (Amati and Meselson, 1965).

In *Aspergillus nidulans* the size of the region is 0.5 map unit (2×10^4 nucleotide pairs) (Pritchard, 1960b).

The frequency of recombination can be assessed in the switching region in various ways. Let us consider the simplest (Chase and Doermann, 1958). Let us determine the frequency of formation of r^+ phages in two four-point crosses (Fig. 31). The frequency of formation of r^+ phages in the first cross (A) may be designated as x, and in the second (B) as y. Let us assume that r^+ appear exclusively on account of recombination events taking place in the switching region. If the interval $b-c$ is not very large, this assumption is perfectly justifiable. In that case it is evident that $2x/(2x + 2y)$ is the probability of recombination in the switching region measured by the interval $b-c$. In Chase and Doermann's experiments the interval $b-c$ varied from 5.6 to 14% and the measured frequency of exchanges in the switching region was 0.3-0.5. The value obtained for the frequency of exchanges was obviously too low, because the intervals used for determination were too long. Stricter measurements (Stahl et al., 1964) give an estimate of 2-3 exchanges per region. In *A. nidulans* (Pritchard, 1960b) its value is 0.6 exchange per region. In short distances (of the order of 1%) very high probabilities of recombination are thus observed in the regions of effective pairing. This means that the scale of the maps is enlarged by tens or hundreds of times in these regions. The situation is similar during transformation and transduction. The dimensions of the genes, determined by transformation or transduction, are expressed not in tenths of a map unit, but in whole map units, i.e., the scale of the map is enlarged tenfold (Carlton, 1966). Within the framework of the effective pairing hypothesis these differences are explained on the assumption that different genetic processes lead to differences in the frequency of contact formation between homologous chromosomes. This frequency must also vary with factors such as the length and structure of the chromosome. The physical scale of gene maps for different organisms may thus vary (1.8). Only the molecular scale within the region of effective pairing must be invariant. The value of the number of nucleotide pairs per map unit within the switching region for *A. nidulans* corrected on this basis, is 150 times smaller

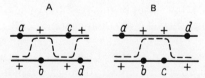

Fig. 31. Determination of frequency of crossing-over in region of effective pairing by comparing frequencies for formation of wild-type recombinants in different four-point crosses.

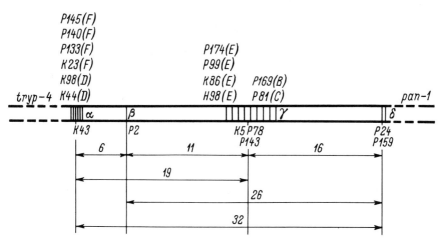

Fig. 32. Map of the *me-2* gene (methionine synthesis) of *Neurospora crassa* (from Murray, 1963): α, β, γ, Δ) groups of mutant sites in the *me-2* gene; *tryp-4* and *pan-1*) genes controlling synthesis of tryptophan and pantothenic acid; $rf \times 10^5$.

than that given in Table 1, namely 270 nucleotide pairs. This estimate agrees well with the figure obtained for phage T4 (Table 1). Pritchard's hypothesis provides a very elegant explanation of high negative interference and it has therefore been widely accepted. However, it is not the only possible explanation and, moreover, it is evidently incorrect.*

2.11. Recombinational Discontinuity of the Chromosomes

The discovery of high negative interference is the most fundamental fact after the possibility of recombination within the gene was proved. Its appearance in interallelic crosses is evidence that intragenic recombinations cannot be explained in terms of the classical concept of crossing-over. A further study of this phenomenon revealed another very important fact: discontinuity of the organization of genetic material with respect to recombination.

During a study of the *me-2* flank system (Fig. 32) a characteristic asymmetry in the distribution of the flank markers was observed (Murray, 1960). In any interallelic cross the R_1 class exceeded R_2 and the P_2 class exceeded P_1 by 4-7 times. Asymmetry of the R classes was understood (2.10), but asymmetry of the P classes was unexpected and unusual. In fact, since the

---

*The *fed⁻* mutants of phage λ, which synthesize products of the Red-pathway in excess, recombine with increased frequency. Consequently, *rf* is limited by the enzyme concentration rather than by the probability of contacts (Franklin, 1971b).

proximal* flank in this system is 1.5 times longer than the distal, and the P_1 class is formed (it is assumed) by additional exchange in the proximal region, it is this class which ought to appear at higher frequency, and not the P_2 class as was actually observed.

In her analysis of these results from the standpoint of the multiple exchange hypothesis in the region of effective pairing, Murray (1963) found that the maximal frequency of additional exchanges in the proximal region is 36%, whereas the minimal frequency of these exchanges in the distal region is 49%. She also found that the frequency of exchanges in the proximal region is a function of the position of the proximal site: the farther the proximal allele from its own end of the gene, the higher the frequency of exchanges in this region.

To explain these facts the hypothesis of fixation of the regions of effective pairing was put forward on the basis of the following assumptions:

1. The chromosome is discontinuous with respect to recombination events. It is built from a series of fixed regions of effective pairing, each of which may be one or several genes in length, and circular (Stahl, 1961) or linear (Murray, 1961). The ends of the fixed regions may coincide with the ends of the genes.
2. During conjugation of homologous chromosomes in meiosis only a small proportion (less than 1%) of these regions participate in effective contacts.
3. Multiple crossings-over are possible in the region of effective pairing.

The closer the site to the end of the fixed region, the lower the probability of recombination at this flank. In the case of *me-2*, the proximal boundary of the region evidently coincides with the proximal end of the gene, while the distal boundary is fixed outside its limits; for this reason P_2 is greater than P_1 (see also 3.6 and Fig. 43).

Asymmetry of the P classes has been found to be a characteristic feature of all flank systems (for a survey, see Whitehouse and Hastings, 1965). It is interesting to note that three genes of *Aspergillus*, namely *ad-9* (Calef, 1957), *paba-1* (Siddiqi and Putrament, 1963), and *ad-8* (Pritchard, 1960a,b), and two genes of *N.crassa*, namely *cys-2* (Stadler and Towe, 1963) and *me-2* (Murray, 1963), are characterized by asymmetry in the same direction relative to the centromere. The suggestion has been made (Siddiqi, 1962) that asymmetry depends on the general organization of the chromosome and that any process essential to recombination is polarized relative to the centromere from the distal to the proximal end of the unfixed region of effective pair-

---

*The term "proximal" means nearer to the centromere. In this case, *tryp-4*. If the locus is not oriented relative to the centromere, its left end is regarded as proximal.

ing. Effective pairing can be interrupted, and this prevents subsequent exchanges. To test this hypothesis the *me-2* region together with its flank markers was displaced from linkage group IV into VI, into the arm opposite to that containing the *me-2* gene, and under these circumstances the distal asymmetry was replaced by proximal (Murray, 1968). Inversion in the same linkage group also was used for reorienting the *me-2* region. The type of polarity was changed in this case also. Consequently, the direction of asymmetry is independent of the general organization of the chromosome, but is determined by the specific properties of the particular locus. This is also shown by the existence of bipolar genes characterized by double asymmetry: *paba-1* and *lys-51* in *Aspergillus nidulans* (Siddiqi and Putrament, 1963; Pees, 1965). Proximal asymmetry ($P_1 > P_2$) is observed in these genes on crossing with each other, while in the distal region the P_2 class is predominant. Opposite asymmetry at different ends of the gene cannot be explained by the hypothesis of multiple exchanges in fixed region of effective pairing (3.6).

2.12. Allele Specificity of Recombination

Hershey (1958) pointed out that the variability of *C*, even in short segments, observed in the experiments of Chase and Doermann (Fig. 28) cannot be explained from the standpoint of the multiple exchange hypothesis. In addition, good additivity (not subsequently confirmed − 2.7) was found in their experiments, and this could not be explained on the basis of Pritchard's hypothesis (narrowing of the map should have been observed). These facts compelled Hershey to propose another explanation of high negative interference, the essence of which was that the presence of the marker itself induces crossing-over. The more markers participate in crossing, the more effective the local stimulation of the recombination processes. In other words, a heterozygous state of the chromosome induces exchanges in this region. In accordance with the model of molecular heterozygosity, this situation appeared to him to be highly probable in phages (5.6), for each marker lying in a heterozygous region forms an additional point of noncomplementarity in it. Similar views were also expressed by Pontecorvo (1958). One of the consequences of this hypothesis is that the degree of high negative interference must correlate with the character of mutation injury. Although no such correlation has been found (Folsome, 1965), facts are accumulating to show that the frequencies of intragenic recombinations may depend on the character of the mutation: allele specificity of recombination.

The first results of this type were obtained by Demerec et al. (1958) for *Salmonella* and by Hotchkiss and Evans (1958) for *Pneumococcus*. A detailed study of this phenomenon was later undertaken by Balbinder (1962).

In the course of genetic analysis of the *tryD* gene (one of the genes controlling tryptophan synthesis) in *Salmonella typhimurium* Balbinder found four exclusive alleles: *D29, D42, D10,* and *D11*. The first two were mapped at the same site and could be reverted to the wild type by the same mutations; consequently, they had injuries of the same character. The *D10* and *D11* alleles also were mapped at the same site, but they evidently differed in the character of their mutation injuries, they differed in their specificity of reversion and their suppressibility.

To match the identity of the *D29* and *D42* alleles observed above, they also behaved in the same way on recombination. Balbinder transduced different *D* alleles into strains *D29* and *D42* carrying the additional marker *tryB4* (Fig. 33), closely linked with *tryD*. For comparison the same alleles were transduced into strains *D7* and *D66*. If *D29* (or *D42*) was crossed with any of the mutants located on its right side on the map (*D55*, for example; Fig. 33A), a definite difference was observed in the frequencies of appearance of the *B4* and *B4$^+$* alleles among the *D$^+$* recombinants, corresponding to the expected value: $\frac{D^+B4}{D^+B4 + D^+B4^+} = 93\%$. If *D29* was crossed with any mutant (*Dx*) on its left-hand side, an abnormally high frequency of recombinations was found in region 3 (Fig. 33B): $\frac{D^+B4}{D^+B4 + D^+B4^+} = 20-60\%$. Because of this abnormal behavior of *D29*, differences were found in reciprocal crosses also. It appeared in fact as though the presence of *D29* stimulated additional exchanges only on its right side. With shortening of region 2 (Fig. 33B) the frequency of stimulated exchanges increased. The aberrant behavior of *D29* and *D42* made it difficult to localize some alleles relative to them. For example, the *D7* allele was on its left side, but for the cross *D29B4* × *D7* the ratio *D$^+$B/D$^+$* = 73%, giving the impression that *D7* was located on the right.

The pair *D10* and *D11* were remarkable in another respect. The *D11* allele behaved perfectly normally in all crosses whereas *D10* behaved aberrantly. These findings show that the recombination was characterized by allele specificity and not by site specificity.

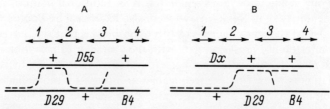

Fig. 33. Crosses in which allele specificity of recombination was found: 1-4) intervals between markers and between markers and ends of the exogenote.

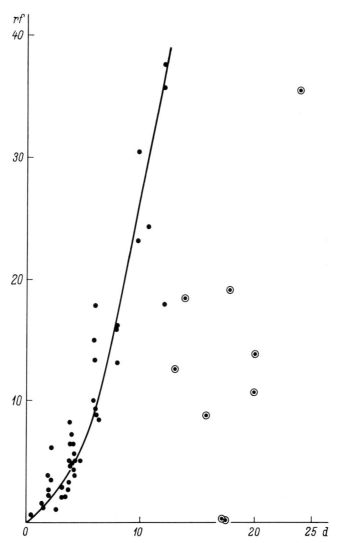

Fig. 34. Effect of allele on the relationship between rf and d (from Smith, 1965). Circles are drawn around points determined in crosses involving $K550$ allele of the his-2 (histidine synthesis) gene of *Neurospora crassa*. Coordinates of the points were determined by the standard method (Fig. 4). $rf \times 10^5$, $d \times 10^5$.

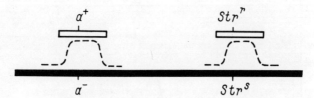

Fig. 35. Determination of integration efficiency (IE) of a marker during transformation. Markers a^+ and Str^r are integrated independently into the chromosome, since they are on different molecules during isolation of DNA from the donor strain. Str^r − reference marker. If a^+-transformants appear in this cross with a frequency of 2×10^{-3}, and Str^r-transformants with a frequency of 1×10^{-2}, IE of the a^+ marker will be 0.2.

Clear examples of allele-specific recombination are described by other workers (Drapeau et al., 1968; Kemper, 1970). The effect of the character of mutation on additivity is very well illustrated by results obtained by Smith (1965) (Fig. 34).

Allele specificity of recombination is clearly demonstrated during the transformation of bacteria. In the course of two-point crosses in transformation experiments *rf* values between markers are standardized relative to the frequency of transformation of a linked or unlinked marker in order to avoid errors associated with the irreproducibility of the physiological conditions. Apparently the *rf* value, when standardized in this way, is a reasonably good indicator of the mutal arrangement of sites on the map. However, attempts to draw maps purely on the basis of *rf* values have frequently given unsatisfactory results (Ephrati-Elizur et al., 1961). Considerable differences in *rf* values have also been observed in reciprocal crosses. The fact is that the integration efficiency (IE) of the markers has a considerable influence on *rf* (Lacks and Hotchkiss, 1960). The IE of the marker is determined in crosses in which the donor and recipient strains differ only in this marker and in the unlinked reference marker (Fig. 35). For example, in the cross $a^+Str^r \times a^-Str^s$, where a^+ is the test marker and a^- its allele, Str is the gene determining resistance (r) or sensitivity (s) to streptomycin, IE is determined as the ratio between the number of a^+ transformants and the number of Str^r transformants. The integration efficiency of the markers can vary within very wide limits. If IE of the reference marker is taken as 1, IE of the test markers can vary from 1 to 0.001 (Lacks, 1966).

Lacks isolated 19 deletions and 57 point mutations in the amylomaltase (*am*) gene of *Pneumococcus*. The gene determining dependence on sulfanilamide (*sulf-d*) was used as the reference marker. He found that the point mutations fell into four discrete classes on the basis of their IE values: 0.026-0.085 (A); 0.17-0.26 (B); 0.29-0.64 (C); 0.86-1.00 (D). Both HNO_2 and

NH$_2$OH were found to induce mutations in class A only, while mutations in class B were induced only by ultraviolet light. For the deletions there was a continuous distribution of IE values over a very wide range (from 0.005 to 0.83). Large deletions were characterized by a high IE. Allele specificity is proved by the fact that two point homoallelic mutations $E7$ and $E02$ possess very different IE values. Furthermore, point mutations in regions occupied by deletions with high IE may possess a low IE. This also shows that neither did the degree of change nor the general structure of the region have any effect on IE. All that matters is the nature of the mutation injury.

Ephrussi-Taylor and collaborators (Ephrussi-Taylor et al., 1965; Sicard and Ephrussi-Taylor, 1965; Gray and Ephrussi-Taylor, 1967) chose a very convenient system for investigation. Mutations in the *ami-A* locus of *Pneumococcus* lead to the appearance of resistance to aminopterin and, at the same time, to sensitivity toward sharp changes in the relative concentrations of isoleucine, leucine, and valine in the medium. The advantage of this system is that a selective medium can be used both for transformation of the mutants to the wild type and for reciprocal crossing. This immediately led to the discovery of the following striking fact: IE of the marker is independent of the direction of crossing. Point mutations at this locus can be divided into only two nonoverlapping classes: integrations of high efficiency, amounting to 0.7-1.6 IE of the standard marker *Str-r41* (HE sites) and of low efficiency (0.07-0.14; LE sites). Deletions showed a continuous distribution of their IE values. All mutations induced by HNO$_2$ and ethylmethanesulfonate fell into the LE class. Mutants of both types and mutants with intermediate IE values appeared spontaneously. The difference between the experimental results obtained by Lacks and Ephrussi-Taylor and his collaborators is attributed to the fact that the hydroxylamine and ethylmethanesulfonate used by Lacks were not in fact mutagenic under his experimental conditions. Evidently the mutations which Lacks regarded as induced were in fact of spontaneous origin. It was shown (Gray and Ephrussi-Taylor, 1967) that only mutations induced by proflavin (deletion or insertion of one or several bases) possess a continuous IE spectrum (from 0.03 to 2), whereas all other mutagens inducing simple or combined substitutions (HNO$_2$, hydroxylamine, nitrosoguanidine) lead to the formation of LE class mutations only. The systematic study of this system enabled some very curious conclusions to be drawn regarding the connection between IE of the markers and their behavior during recombination. To begin with the findings of Lacks and Hotchkiss (1960), that IE of the marker in the recipient chromosome has a much greater influence on *rf* in two-point crosses than the distance between them and the donor marker, were confirmed. Accordingly sharp differences arise in reciprocal crosses. For example, if an *r6* (LE) recipient is crossed with an *r22* (HE) donor, *rf* = 0.020%. In the reciprocal cross (recipient *r22*, donor *r6*) *rf* = 0.18-0.52% (Sicard and Ephrussi-Taylor, 1965). To avoid the diffi-

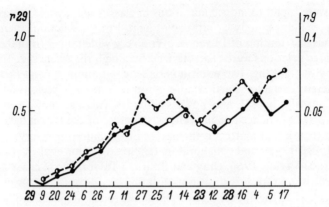

Fig. 36. Effect of integration efficiency of the marker on the scale of the genetic amp during transformation of *Pneumococcus* (After Gray and Ephrussi-Taylor, 1967). The abscissa is the map of the *ami-A* locus on which only the order of the mutant sites is shown. The reference numbers of the HE alleles are shown in bold type. Values of *rf* determined in crosses of the corresponding alleles with the HE allele *r29* in the recipient strain are plotted along the left ordinate. The broken line joins points whose ordinates were obtained in crosses of this type. Values of *rf* obtained in crosses with the LE allele *r9* are plotted along the right ordinate. Ordinates of points joined by the continuous line were determined in crosses of this type.

culties associated with this effect when mapping the *ami-A* locus two series of crosses were performed: in one series the recipient was always a definite LE marker (*r9*), and in the other it was an HE marker (*r29*). Any markers were used as donors. On the basis of *rf* values obtained within each series good gene maps with satisfactory additivity and an identical arrangement of sites can be drawn. However, the scales of these maps differ by as much as ten times (Fig. 36).

Later work showed that the probability of integration of the HE site in *cis*-2-site* crosses involving markers with different IE values is lowered by the presence of a linked LE site (exclusion effect). Lacks showed that the degree of exclusion of the HE site depends on the distance between the markers (it is inversely proportional to the distance). Gray and Ephrussi-Taylor confirmed and extended these observations. They found that the degree of exclusion in the *ami-A* locus also depends on the orientation of the HE site relative to the LE site (the polarity of exclusion). If the LE site lies on the right of the HE site the exclusion is stronger than if it is on its left.

*Mutant alleles on the same chromosome ($ab \times a^+b^+$).

Attempts to study the nature of various HE and LE alleles (Ephrussi-Taylor et al., 1965; Noubo et al., 1966) led to a curious discovery. The LE markers were found to be highly sensitive to ultraviolet irradiation whereas the HE markers were resistant (i.e., they are inactivated at a much slower rate) (Patrick and Rupert, 1967; Munakata and Ikeda, 1969). It was also found that during transformation by complementary DNA strands the IE of the markers varies depending on which DNA strand (right or left) the marker lies. In this case, *rf* also varies (Gabor and Hotchkiss, 1966).

Later we shall study in detail the explanation of these various anomalies.

2.13. Conclusion

The discovery of interallelic recombination led to a radical revision of classical ideas of the gene and of genetic recombination.

1. It was found that the gene has a complex structure and consists of hundreds of units (sites), arranged in linear order and capable of undergoing mutation and recombination independently. The linear character of gene maps has been established by several independent methods: the method of overlapping deletions, the method of two-point and three-point crosses, and the method of flank markers. That the genetic picture of gene structure corresponds adequately to the physical picture is proved not only by the fact that its physical basis is the linear sequence of nucleotides in the DNA molecule, but also by the colinearity of the gene and the polypeptide chain which it codes, or in other words by the fact that the position of the mutant site on the gene map corresponds to the position of the modified amino acid component in the polypeptide (Sarabhai et al., 1964; Yanofsky et al., 1964). This also indicates that intragenic *rf* values are approximately proportional to physical distance (i.e., to the number of nucleotide pairs between the sites).

2. The limit of divisibility of the gene is one nucleotide, as is shown by the possibility of intracodon recombination (i.e., recombination within the triplet of nucleotides coding a particular amino acid). However, the recombination frequencies at the level of a few nucleotides were found to be ten times less than those expected theoretically on the basis of the assumption that the physical scale of the gene map is constant. This leads to considerable "widening" of the genetic map.

3. A unique feature of intragenic recombination is its allele specificity. The recombination frequency between alleles is very considerably affected by the nature of the mutation injury. Allele specificity of recombination is most clearly manifested in reciprocal crosses during transformation. As the work of Ephrussi-Taylor shows, aberrant behavior of alleles during crossing is manifested only if the alleles taking part in the cross differ in their IE values.

This fact explains why the largest number of aberrantly recombining alleles and the poorest quality of intragenic maps are obtained by the use of spontaneous mutations. If homogenous groups of mutations induced by the same mutagen are mapped anomalies are found much more rarely, although in every paper describing the mapping of genes it is stated that some of the mutants had to be discarded. For strict accuracy the gene must therefore be mapped by at least two independent methods.

4. A characteristic feature of intragenic recombination is high negative interference, manifested as an excess of P and R_2 classes in the standard flank system and an excess of double recombinants in three-point intragenic crosses. To account for this phenomenon Pritchard put forward the hypothesis of effective pairing and postulated continuous synapsis between homologues and a very high recombination frequency on short segments in the regions of effective pairing. The presence of asymmetry in the distribution of the P classes led to modification of Pritchard's hypothesis so that the regions of effective pairing came to be regarded as fixed on the chromosome.

Besides effective pairing, two other hypotheses have been put forward to explain high negative interference: the hypothesis of marker stimulation and the conversion hypothesis. Special experiments failed to prove the first of these, although allele specificity of recombination was found; the second was of very great heuristic value and is evidently the closest approximation to the truth. This hypothesis is discussed within the framework of general analysis of the problem of gene conversion (Chapter 3).

CHAPTER 3

Gene Conversion. Tetrad Analysis of Intragenic Recombination

3.1. History of the Problem

About a dozen different names have been suggested for the phenomenon which here and later in the book will be called conversion*: the abundance of synonyms clearly reflects the atmosphere of speculation which until recently has enveloped this phenomenon. For various reasons, however, none of the newly introduced terms gained wide acceptance and recognition, and in modern genetic literature the classical name — gene (or allelic) conversion — is still applied to this phenomenon.

Without knowing the mechanism of gene conversion we can only define it provisionally as a phenomenon revealed by deviations of various types from 1 : 1 Mendelian segregation in tetrads. If two homologous chromosomes, one marked with the allele a and the other with the allele a^+ are united in the zygote of a haploid fungus, during meiosis the chromosomes segregate into different nuclei. On analysis of the tetrad formed a regular segregation is found with respect to the character determined by this gene — two a spores : two a^+ spores, or in organisms with a sac containing eight spores, two a pairs : two a^+ pairs (Fig. 37A). If a mutation in the gene determining color of the ascospores is used as the genetic marker, the regularity of this type of segregation can easily be observed in tens and hundreds of tetrads. With a sufficiently wide range of selection of tetrads, however, asci with an aberrant spore ratio are found — $3a^+ : 1a$ and $1a^+ : 3a$ (Fig. 37B,C) or, more rarely still, $4a^+ : 0a$ and $0a^+ : 4a$. Other types of aberrant segrega-

---
*Nonreciprocal recombination (Mitchell, 1955); the Mitchell effect (Roman, 1956); transmutation (Horowitz, 1957), transreplication (Glass, 1957), intragenic recombination (Stadler, 1959a), defective copying (Pritchard, 1960a), reversion (Magni and Von Borstel, 1962), heteroallelic regeneration (Hurst and Fogel, 1964), asymmetrical crossing-over (Westergaard, 1964).

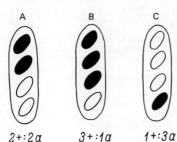

Fig. 37. Normal (A) and aberrant (B and C) tetrads obtained by crossing mutant a with the wild type (a^+). Mutation in the a gene leads to depigmentation of the spores, colored black in the wild type.

tion in tetrads, discovered only very recently, will be examined below (3.9). Such deviations from the expected 1 : 1 segregation were first discovered by Kniep (1928), the pioneer of tetrad analysis, and soon after by many other investigations (for a survey of the literature for this period, see Zickler, 1934).

The attempt might be made to explain segregation of the 3 : 1 type by mutation of one of the alleles at the four-chromatid stage, and the appearance of 4 : 0 tetrads by the occurrence of mutation at the chromosome level, i.e., before it splits into two chromatids. However, the frequency of spontaneous mutations is too low ($10^{-6}-10^{-8}$) to account for the fairly high probability of discovery of such tetrads (up to 10^{-2}). Either the frequency of mutations is sharply increased in the heterozygote or conversion of one allele into another takes place as the result of some other event.

The term "conversion" was introduced into the literature on genetics by Winkler (1930, 1932) to describe such cases of aberrant segregation in tetrads. He regarded conversion as a direct mutation of one allele into another induced by the heterozygous state of the locus. On the basis of these concepts Winkler developed his conversion hypothesis of recombination of linked genes, an alternative to the hypothesis of crossing-over. According to his views, no exchange of material takes place between the chromosomes, but any allele taking part in crossing can be converted into the opposite allele (Fig. 38). Winkler distinguished between equal (the interconversion of

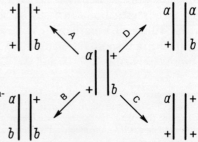

Fig. 38. The conversion mechanism of recombination of linked genes. It results in the absence of a reciprocal recombinant. No physical exchange likewise takes place between the homologous chromosomes. Each line represents a chromosome. A) conversion $a \to +$; B) conversion $+ \to b$; C) conversion $b \to +$; D) conversion $+ \to a$.

alleles takes place at equal frequency) and unequal (with unequal frequency) conversions as well as monogenic (conversion in one locus in any direction) and digenic conversion (in both loci in different directions).

Unfortunately the attempt to describe recombination of linked genes on the basis of the view of gene conversions played a fatal role in the history of the study of this most interesting phenomenon. After it had been convincingly shown that recombination of linked genes is accompanied by cytologically observable exchanges of large segments of homologous chromosomes (1.12), the hypothesis of crossing-over was converted into a precise scientific theory, and Winkler's ideas, and with them the problem of gene conversion, were forgotten for a very long time. As regards the cases of aberrant segregation in tetrads, they were explained in terms of the dominant theory as cytological anomalies (heteroploidy, etc.) (Roman et al., 1951).

Attention was drawn back to this problem only after Lindegren (1949), working with yeast, had observed the appearance of two unusual tetrads among the large number of normal tetrads obtained from a white X pink (alleles of the gene determining the color of the colonies) cross. Each tetrad contained three pink and one white spore. Four years later (Lindegren, 1953) another tetrad of the same type was found from a total number of 2500 tetrads. Other workers (Mundkur, 1949) also described the discovery of various types of abnormal spore ratios in the tetrads. Lindegren considered that aberrant segregation in tetrads can be explained by the conversion postulated by Winkler. He asserted that conversion exists as a mechanism of genetic variation which differs in principle from crossing-over. Lindegren gave this phenomenon the following definition: "gene conversion is interaction between dominant and recessive alleles in meiosis as a result of which the dominant allele is transformed into the recessive, or vice versa" (Lindegren, 1955, p. 605).

Lindegren's contribution was that, by his work and his unswerving conviction, he once again drew the attention of investigators to the forgotten phenomenon of conversion. However, not even he could prove beyond all doubt that conversion really exists, for the system with which he worked had no closely linked markers. For this reason the appearance of aberrant asci could have been due to a variety of cytological anomalies or even to elementary contamination (Strickland, 1958a), and a suitable control was necessary. Such a control could be provided in a system with closely linked markers. If conversions in different genes take place independently, in a digenic cross it would be possible to find regular segregation for one character and aberrant for another. The discovery of such tetrads would completely rule out the possibility that the aberrant spore ratios in the tetrads could be due to cytological or genetic anomalies of any type, for in these cases aberrant segregation would necessarily be observed for both closely linked markers simultaneously. Markers of this type were used by Mundkur

(1949) in his investigations, and it was found that in asci with an aberrant segregation for the *Mel* gene (determining ability to ferment melibiose) the closely linked *Gal* marker (fermentation of galactose) nearly always segregated normally. However, even these results did not rule out the possibility of other interpretations connected with polygenic determination of sugar fermentation characters in yeast (for a survey of the literature of this period, see Emerson, 1956).

3.2. The Conversion Hypothesis of High Negative Interference

Although most geneticists were highly skeptical of gene conversion, Mitchell (1955, 1956) found this idea attractive for it could provide a basis for the interpretation of the unusual distribution of flank markers which had been found without the need for invoking the hypothesis of multiple exchanges on short segments (Giles, 1951). If wild-type recombinants are formed during crossing of allelic mutants not as a result of crossing-over between them, but through frequent conversions of these alleles to the wild type, this can explain the excess of P and R_2 classes observed in flank systems (2.8) and which can be described in terms of interference. The P_1 class in the standard flank system (Fig. 26) arises directly as the result of conversion of the proximal *a* allele to the wild type (Fig. 39), while P_2 appears through conversion of the distal *b* allele. Exchanges at the flanks are essential for the appearance of R classes and, consequently, the conversions must be somehow connected with crossing-over.

In order to explain the nature of high negative interference it was thus necessary to proceed by tetrad analysis. If wild-type recombinants arise through crossing-over between alleles, double mutants — complementary recombination products — must be found in tetrads containing these spores (of wild type). If no such mutants are found, this will be an argument in

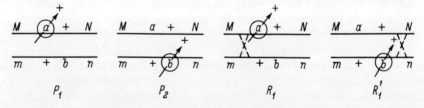

Fig. 39. Formation of different configurations of flank markers in a standard flank system (Fig. 26) according to Mitchell's hypothesis. Converting alleles are circled. Arrows indicate direction of conversion. P_1) Conversion $a \to +$; P_2) conversion $b \to +$; R_1) conversion $a \to +$, accompanied by crossing-over in the proximal region; R_1') conversion $b \to +$ and exchange in the distal region. Exchanges marked by broken lines.

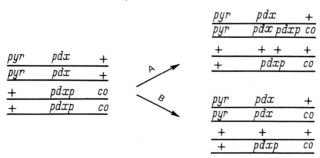

Fig. 40. Formation of asci of different types in *Neurospora crassa* during crossing-over between alleles *pdx* and *pdxp* (A) and conversion of the *pdxp* allele to the wild type (B).

support both of conversion and of the conversion hypothesis of high negative interference. The laborious analysis was undertaken and this classical work by Mitchell opened a new chapter in the study of the mechanism of genetic recombination.

3.3. Mitchell's Discovery

Mitchell (1955) crossed strains of *Neurospora crassa* marked by different alleles of the *pdx-1* locus (controlling pyridoxin synthesis) and by flank genes *pyr-1* (pyrimidine synthesis) and *co* (colonial growth of mycelium): *pyr pdx* × *pdxp co* (Fig. 40), where *pdx* and *pdxp* are mutant alleles of the *pdx-1* gene. Of a total of 585 asci she found only four which contained wild-type recombinants (with one pair of wild-type spores in each ascus). These asci could be regarded as tetratypes (1.9) appearing as the result of crossing-over between alleles (Fig. 40A). To test this hypothesis it was necessary to undertake the genetic analysis of mutant spores from these asci and to show that one of the pairs of spores is a double mutant *pdx pdxp* (the reciprocal product of crossing-over). One member of each sister pair (it was assumed that the sister spores are identical, for they were formed by mitotic division — see 3.9) was crossed with the two parent strains. Evidently the double mutant does not form wild-type recombinants in either of the crosses analyzed. If, however, the ascus does not contain this double mutant, all the mutant spores will give wild-type recombinants when crossed with either parent. It was this last result which was obtained when all four asci were analyzed (Fig. 40B). The results showed that one of the mutant alleles (in this example, *pdx*) is represented in the normal number in asci containing wild-type spores (i.e., its segregation is 1 : 1), whereas the other (*pdxp*) is present in a smaller number (one pair of spores of this genotype instead of two, segregation 3+ : 1*pdxp*). Since the flank markers (and, in particular, the second allele) segregated normally, the evidence that asci with aberrant segregation appear, not as the result of cytological anomalies, but as the result of conversion, was absolutely convincing.

Mitchell's work not only finally convinced everyone that conversion really exists, but also revealed the genetic importance of this phenomenon. The formation of wild-type recombinants in interallelic crosses is not necessarily accompanied by the formation of double mutants, i.e., intragenic recombination may be nonreciprocal and can take place by a mechanism (conversion) which differs sharply from that of the classical crossing-over.

3.4. Crossing-over or Conversion?

After the appearance of Mitchell's work it was no longer possible to deny that conversion exists, but Mitchell's interpretation of the genetic significance of this phenomenon, and the conversion hypothesis of high negative interference linked with it, aroused criticism from a number of geneticists (Roper and Pritchard, 1955; Pontecorvo, 1958). They claimed that although conversion takes place it is only an accidental "copying error" (to use the terms of the copy-choice hypothesis, see 3.5), occurring at low frequency and not a regular mechanism of recombination of alleles unlike crossing-over, the natural mechanism of recombination at both intergenic and intragenic levels. As support for their views, the defenders of the hypothesis of a single reciprocal mechanism cited the possibility of finding reciprocal recombination products in interallelic crosses (2.3). However, these citations referred to the analysis of random choices in the progeny and they were therefore not convincing. In fact, the phenomena observed during conversion can easily be confused with crossing-over if conversion can take place not only from mutant to wild type ($a \rightarrow +$), but also from wild type to mutant ($+ \rightarrow a$). In this case recombination products, simulating reciprocity, can be found by statistical analysis. However, only the use of tetrad analysis can show whether they appear as the result of one or of different events (i.e., whether they are in one or different asci). Roman (1958), followed by other investigators (Rizet and Rossignol, 1963; Case and Giles, 1964; Fogel and Mortimer, 1968), showed that double mutations can arise as the result of conversion, i.e., without accompanying wild-type recombinants in the same ascus. Their work also showed that the absence of a double mutant in asci containing wild-type spores is the result of the recombination mechanism itself (and is not due, for example, to its nonviability).

Another argument was based on the experimental results obtained by Roper and Pritchard (Roper and Pritchard, 1955; Pritchard, 1960b) by semitetrad (Fig. 41) analysis of mitotic intragenic recombination in *Aspergillus*. It takes place spontaneously at a very low frequency, only slightly above the "noise" level. Nowadays inducing agents are used to increase its fre-

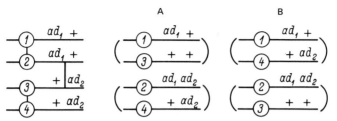

Fig. 41. Mitotic recombination between alleles ad_1 and ad_2 (adenine synthesis) in *Aspergillus nidulans*. Each line on the left-hand side of the diagram represents a chromatid, and on the right-hand side a chromosome. A and B are alternative forms of separation of the chromatids (1-4) in the daughter cells (between parentheses).

quency (1.16). At that time no such methods had yet been developed, and Pritchard therefore selected mutants exhibiting the highest percentage of recombinations with each other for semitetrad analysis of crosses involving alleles of the ad_8 gene (one of the genes controlling adenine synthesis). If two mutant alleles of the ad_8 gene coexist in a diploid cell (heteroallelic diploid), as a result of mitotic recombination ad^+ prototrophs appeared and, as before, they remained diploid (Fig. 41), i.e., they contained two of the four chromatids participating in the recombination (semitetrad). The diploid prototrophs were then analyzed to see if they contained a complementary recombinant chromosome. Under selective conditions only cells containing the third chromatid (chromosome) are recorded. No reciprocal product can thus be found in 50% of cases. The more precise figure is in 77% of cases (allowing for crossovers between the gene and centromere; Raipulis and Leh Dinh Lyong, 1969). By 1960 Pritchard had analyzed 52 ad^+ diploids. Of these, in only 12 (23%) cases were cells still diheterozygous for these alleles, but they were now in the *cis* position, i.e., they were evidently formed by a reciprocal event. However, since the frequency of appearance of ad^+ in Pritchard's experiments was low enough to be compared with the frequency of back mutations, some cases of the observed nonreciprocity were ascribed by him to the mutation pool. Pritchard concluded that recombination within the gene is predominantly, if not always, reciprocal in character, i.e., that it takes place on account of crossing-over.

Later in the book asci (tetrads or octads) containing reciprocal recombination products will be designated by the letter t,* while those not containing such products, i.e., formed by conversion, will be designated by the letter c.

Since the relationship between crossing-over and conversion in intragenic recombinations was a problem of the utmost importance, its elucida-

---

*t refers to the tetratype in interallelic crosses, by contrast to T, which is the tetratype in intergenic crosses (1.9).

tion has been the target of much research. However, results have accumulated very slowly and in many cases they were not statistically convincing. This was because mainly biochemical mutants have been studied and the tetrad analysis of crosses in such cases is a laborious task.

The application of semitetrad analysis to yeast has shown that mitotic intragenic recombination in yeast is mainly nonreciprocal in character. Roman and Jacob (1958) found not a single t type among 52 recombinant semitetrads obtained by crossing alleles *1* and *2* of the *is-1* gene (isoleucine synthesis). These results were confirmed by Kakar (1963) who found 29 c semitetrads in 29 recombinants for the same alleles of the *is-1* gene, and also by Roman (1963), who developed a technique of complete detection of all four products of mitotic recombination between alleles of the ad_6 gene. The essence of this method is as follows.

A diploid homozygous for ad_2 is formed. The cells of this genotype cannot grow in the absence of adenine, but on medium with adenine they form red colonies. If such a diploid is also heteroallelic for another gene ad_6, the colonies grow white. In the process of colony formation mitotic recombination between the ad_6 alleles can take place in some cells, as a result of which these cells will give rise to subclones visible as red sectors in a white colony. Genetic analysis of the red and white sectors reveals all the viable recombination products. More than 100 tetrads were analyzed by this method and none was found to belong to the t type. Similar results were obtained with the same system by Raipulis and Leh Dinh Lyong (1969).

During the study of meiotic recombination in *N. crassa* by various workers altogether 132 recombinant asci obtained from interallelic crosses in different parts of the chromosomes were analyzed. Of this number only eight contained reciprocal recombinants (see the survey by Emerson, 1966b). Because of the extreme rarity of discovery of tetrads of this type in meiosis in *N. crassa* and in mitosis in yeast, a more careful analysis of mitotic intragenic recombination in *Aspergillus* had to be undertaken (Putrament, 1964). Numerous *paba-1* alleles (synthesis of *p*-aminobenzoic acid) were used. Among 393 recombinant semitetrads obtained from different crosses there were only 20 of the t type (5%). In crosses of ad_9 alleles (Martin-Smith, 1961, cited by Whitehouse and Hastings, 1965) 15% of t semitetrads (among 126) were found. The t semitetrads appeared only in crosses in which mutant *13* took part. This showed that allele-specificity is possible in the formation of t tetrads.

As a whole, although it was deduced from them that recombinations between alleles take place mainly as a result of conversions, the results obtained by analysis of crosses involving auxotrophic mutants were too few to give a clear understanding of the relationships between reciprocal and nonreciprocal recombinations in interallelic crosses. Progress toward a solution of this problem only became possible more recently through the use of mutants affecting ascospore pigmentation and through an increase in the resolving power of tetrad analysis for the study of biochemical mutants. It has already been stated that the use of mutants for ascospore pigmentation makes the search for recombinant asci much easier. In the discomycete *Ascobulus immersus,* for instance, the intact asci scattered from the fruiting body can easily be collected in any quantity on the surface of the agar and those containing wild-type spores (dark brown; mutant spores are white or pink) can be found visually under the microscope. Unfortunately testing for allelism in this system is difficult. In early investigations (Lissouba et al., 1962) the mutations were therefore distributed not by genes, but by series, on the basis of the recombination criterion. Within each series mutations are closely linked; mutations belonging to different series are linked only weakly or not at all. We now know that each series corresponds to one gene (see page 69; Paszewski and Surzycki, 1964; Gajewski et al., 1968; Celis et al., 1973; Rossignol, 1969). In the course of analysis of this system genes were found in which recombinants were formed mainly as the result of conversions. For example (Lissouba, 1960; Lissouba et al., 1962, some of these data are given below in Table 4) 29 crosses were carried out involving 13 mutants of series 46 and 250 octads (from 1 to 35 in each cross) were analyzed; in only 6 of them were reciprocal products of recombination (t octads) found, and these asci were obtained only in crosses in which mutant 277 took part (allele specificity). However, it would be premature to con-

TABLE 4. Genetic Structure of 6−: 2+ Asci Obtained from Crosses between Mutants of Series *46* (Lissouba et al., 1962)

Crosses $a \times b$	6−: 2+ asci	t asci	c asci	
			$c\ (a \to +)$	$c\ (b \to +)$
188 × 63	8	0	0	8
188 × 46	2	0	0	2
188 × w	10	0	0	10
188 × 138	11	0	0	11
63 × 46	2	0	0	2
63 × w	18	0	0	18
63 × 1216	5	1	0	4
63 × 138	35	0	0	35
46 × w	7	0	0	7
46 × 138	22	0	0	22
w × 1216	18	0	0	18
w × 138	15	0	0	15
1216 × 137	2	0	0	2
Total . . .	155	1	0	154

clude (as these investigators did) from these results that intragenic recombination in this region is entirely nonreciprocal. In fact, the ranges of choice in each cross were small. If the range of choice was doubled (32 octads) one t ascus was found in one of the crosses (not involving mutant *277*; Rizet and Rossingnol, 1963). A similar situation was observed during crossing of alleles of the *y* gene (Gajewski et al., 1966). Of 326 recombinant octads in this case only 18 of the t type were found, 6 of them in crosses involving the *y* allele and the other 12 appeared about equally often (~5%) in nearly all crosses. Solid confirmation of this ratio between the t and c tetrads in intragenic crosses is provided by the extensive statistical material obtained for four alleles of the *hi-1* gene (histidine synthesis) in yeast (Fogel and Hurst, 1967): t tetrads appeared at relatively low frequency (less than 10%) in all crosses (of 1081 asci 101 were of the t type).

More complex relationships between the relative frequencies of appearance of c and t asci were found in other genes and their detailed analysis required certain concrete hypotheses (Chapter 5). However two main tendencies are revealed by even superficial analysis: 1) the frequency of appearance of the t tetrads is directly proportional to *rf* and 2) allele specificity of appearance of the t tetrads. For example, in series 19 (Lissouba et al., 1962) in all crosses (except allele *55*) 34 t and 127 c octads were found, whereas in crosses involving allele *55* there were .160 t and 177 c octads. Alleles giving a high percentage of t octads often behave unusually in another respect: they show postmeiotic segregation (3.9). This tendency is well illustrated by the observations of Fields and Olive (1967): all five alleles of the *hy* gene of *Sordaria brevicollis* give a high percentage (over 25) of t octads and postmeiotic segregation.

The nonreciprocity of intragenic recombinations, like high negative interference, is not a unique feature confined to the objects listed above. In every case in which (because tetrad analysis was impossible) semitetrad analysis of intragenic recombination has been carried out (in *Ustilago*: Holliday, 1966a; in *Drosophila*: Hexter, 1963; Smith et al., 1970; Chovnick et al., 1970, 1971; Carlson, 1971; in flowering plants: Hagemann, 1958; Renner, 1959; Hagemann and Snoad, 1971; Abel, 1971; in *Escherichia coli*: Herman, 1968a; Berg and Gallant, 1971), nonreciprocity (conversion) also was recorded.

Recombination between genes in phage f_1 of *E. coli* is also entirely nonreciprocal (Boon and Zinder, 1971). This has been shown by the use of flank markers and alleles of the gene controlling DNA synthesis (so that recombination took place as far as replication). The content of individual yields corresponded exactly to that expected in the case of gene conversion (Fig. 39: in case P_1, besides wild-type recombinants, the yield also contained phages with the P_2 genotype, while in case P_2, P_1 phages were present. Class R_1 was not observed, for crossing-over led to the formation of a double ring;

3.5. The Switch Hypothesis

During the study of intragenic recombination by tetrad analysis it became increasingly clear that nonreciprocal events are not accidental and that conversion is the predominant, and perhaps the only, mechanism of recombination between alleles. However, were these facts conclusive evidence for the conversion hypothesis of high negative interference? Indeed not. This hypothesis could be proved only by the discovery that conversion and crossing-over are actually fundamentally different mechanisms. Mitchell's hypothesis was based to some extent on the view, handed down from Winkler, that conversion is a special type of mutation. The first steps toward elucidation of the nature of conversion were taken in this direction. For instance, Pritchard argued that conversion cannot be a mutation event, for the frequencies of conversions of crossed alleles increase as the distance between them increases, i.e., the frequency of the intragenic recombination is a function of distance. Consequently, conversion is primarily a recombination event. Furthermore, if conversion is purely the result of an increase in the frequency of mutations in the heterozygote, the converted alleles in the overwhelming majority of cases cannot be genetically identical with the parental alleles, for mutations may be of different types and may arise at different sites of the gene. It was shown (Mitchell, 1955; Roman, 1956; Case and Giles, 1958), however, that the converted alleles do not differ from the parental alleles with respect either to recombination, to mutation, or to complementation. Wild-type spores in the c ascus (3+ : 1−) are indistinguishable from each other at the gene product level also (Zimmerman, 1968). During conversion the information contained in the homologous chromatid is reproduced with an accuracy of one nucleotide (Fogel and Mortimer, 1968, 1970). (For the rare exceptions to this rule, see Emerson and Yu-Sun, 1967; Kitani and Olive, 1967). Another manifestation of the difference between conversions and mutations is that even nonreverting mutations (of the deletion type) can convert. This fact is decisive evidence in support of the hypothesis (Mitchell, 1955, 1956; Lederberg, 1955; Freese, 1957) that genetic information for the extra allele is supplied by the homologous chromosome during DNA synthesis.

It was then already clear that genes are segments of the DNA molecule which differ from each other only in their base sequence, and it could accordingly be expected that recombination is the same process whether between genes or within the gene. On the basis of these considerations Freese (1957) put forward the switch hypothesis, according to which the same

Region of effective pairing

Fig. 42. The mechanism of genetic recombination (after Freese, 1957). Continuous lines represent parental, and broken lines daughter DNA molecules (chromosomes). Arrows indicate direction of replication. "Switching" of the replica can take place within the region of effective pairing.

event (switching of the copying process) can lead to both conversion and crossing-over. Freese saved the hypothesis of effective pairing by combining it with the mechanism of copy-choice. Frequent exchanges within Pritchard's regions were due not to a mechanism of breakage and reunion, but to a copy-choice mechanism. For the reasons described above (1.19) it was necessary for him to postulate conservative synthesis of DNA. In Freese's terminology the original homologous chromosomes are called "parent" and the newly synthesized chromosomes "daughter." Each of the two daughter chromosomes replicates along its own template until it reaches the "switching region" (probably Pritchard's region of effective pairing). In this region replication can switch to the corresponding point on the other template. Having travelled a short distance along the new template the replica leaves it and switches back to the original template. These switches of the replica can take place repeatedly in a short segment of chromosome. The second replica is not bound to repeat all the switches of the first and it leaves the switching region on the free template (Fig. 42). This mechanism permits surplus replication of short segments of the chromosome (which leads to conversion) and at the same time, it permits crossing-over, i.e., recombination of the flank markers.

3.6. The Polaron Hypothesis

One result of the high frequency of switches postulated by Freese is equality of the P_1 and P_2 configurations of the flanks in convertants without regard to the position of the mutant sites. After the discovery (2.11) that this condition is not satisfied, Stadler (1959a) suggested that the observed asymmetry of the P classes may be due to differences in size of the mutation injuries in the crossed mutants. The finer details of this hypothesis were subsequently worked out as applied to transformation (Ephrussi-Taylor, 1961). One obvious difficulty which it encounters is that asymmetry is also found when mutants carrying point injuries are crossed.

Stahl (1961) and Murray (1961) postulated that the switching regions are fixed on the chromosome and are not distributed along it at random (2.11). The distribution of the P classes must depend on the position of the sites relative to the boundaries of the switching region (Fig. 43). In the first case (Fig. 43A) the probability of switching between the proximal boundary of the region and site a is less than between site b and the distal boundary, hence $P_2 > P_1$. In the second case (Fig. 43B), it is easy to see that $P_1 > P_2$.

The hypothesis of multiple exchanges, which was originally developed to explain high negative interference, as a result of certain modifications which left its essentials untouched, was thus able also to explain the non-reciprocity discovered in tetrad analysis and the asymmetry of the flanges discovered during statistical analysis. However, it proved impossible to explain the polarity of the conversions within the framework of this hypothesis.

Analysis of intraserial crosses in *Ascobolus* revealed a curious fact (Lissouba and Rizet, 1960; Lissouba et al., 1962) which is clearly illustrated by Table 4 (the two last columns). Recombinant c asci appear exclusively as the result of conversion of the distal site (a map of series *46* is shown in Fig. 88). For example if mutant *188* (the left-hand end of the series) was crossed with any of the mutants of this series, this mutant did not convert in any of the asci which were analyzed: it was always the second mutant, situated on its right, which converted. Conversely, in all crosses involving site *137* (the right-hand end of the series) it was predominantly this site which converted. Sites in the middle of the series converted only in crosses with mutations located on their left; during crossing with sites on their right they did not convert. This effect was called polarity, and the region of genetic material within which polarity is observed was called the polaron (Lissouba and Rizet, 1960).

In series *19* (Lissouba et al., 1962) bipolarity was found, i.e., polarity of opposite direction at different ends of the series. Bipolarity of recombination was shown by the fact that on crossing of mutants located at the

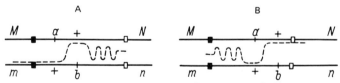

Fig. 43. Diagram showing how fixation of the regions of effective pairing on the chromosome can lead to asymmetry of the flanks. A) The allele a is closer to the proximal boundary of the region (black square) than allele b is to the distal boundary (white square). The probability of switching in the distal region is thus greater and, consequently, class $P_2 > P_1$. B) The probability of an additional switch in the proximal region is higher and, consequently, class $P_1 > P_2$.

left-hand end of the gene (group A) with mutants located in the center (group B) conversion was found mainly in the left-hand site (in 43 of 51 cases), whereas in $B \times C$ (the distal group of sites) crosses mainly the right-hand (C) sites converted (in 40 of 41 cases). Polarity is stronger in the right-hand part of this series because, if mutants belonging to groups with different polarity are crossed, the right-hand mutant converts more often than the left (in 215 of 225 c tetrads). Similar results were obtained by analysis of series y (Kruszewska and Gajewski, 1967). In two-point crosses tetrads containing wild-type spores also are formed mainly as a result of conversion of alleles located on the right on the map. However, the opposite polarity is found in the left half of this gene. It can be concluded that this gene also is bipolar (5.6).

Results obtained on other systems show that polarity is not a unique property of intragenic recombination in *Ascobolus immersus*. The same sort of phenomenon is also found in the *hi-1* gene in yeast (Fogel and Hurst, 1967), in the *6A* region in *Sordaria brevicollis* (Fields and Olive, 1967), and in the *cys* gene of *N. crassa* (Stadler and Towe, 1963). The existence of polarity is also confirmed by the presence of a gradient of the basic frequencies of conversions.

Basic frequencies of conversions (Kruszewska and Gajewski, 1967) is the term given to the frequencies of appearance of aberrant asci in one-point crosses ($a \times +$). The basic frequency of conversion of the mutant a allele to the wild type is designated $fk_{(a \to +)}$, and that of the wild-type allele to the mutant by $fk_{(+ \to a)}$. The total basic frequency $fk_a = fk_{(a \to +)} + fk_{(+ \to a)}$. The first experiments to study conversion showed that $fk_{(+ \to a)}$ and $fk_{(a \to +)}$ often are not equal to one another.

It was also discovered that the values of fk can differ sharply not only for mutant in different genes, but also for alleles of the same gene. Systematic observations on fk values of mutants belonging to the same series were carried out on *A. immersus*. The results relating to series *164* are given in Fig. 44. They indicate the existence of well-defined site-specificity of fk_a and allele specificity of $fk_{(+ \to a)}$ and $fk_{(a \to +)}$. Clearly the allele specificity of fk can often lead to distortion of the gradient (Rossignol, 1969).

Polarization of recombination events within the gene gives the key to the understanding of the character of distribution of flank markers in the standard flank system. If recombination within the gene takes place mainly as a result of conversion of only one of the sites, one of the P types will be predominant. The extensive data recently obtained for the *hi-1* flank system in yeast (Fogel and Hurst, 1967) are in full agreement with this interpretation.

Let us consider the possible alternatives for the formation of recombinant tetrads in an $a \times b$ cross involving flank markers M and N ($Ma+N \times m+bn$). As Table 5 shows, depending on which events take place inside the

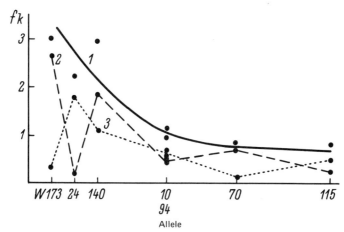

Fig. 44. Gradient of basic frequencies of conversions (fk) for different alleles of series *164* of *Ascobolus immersus*. (Plotted from data of Baranowska, 1970). The abscissa is the map of the series; $fk \times 10^3$ plotted along the ordinate. 1) fk_a; 2) $fk(+ \to a)$; 3) $fk(a \to +)$.

gene and which on its flanks, 12 basic types of tetrads containing wild-type spores can be formed.

Until recently very few results of the tetrad analysis of flank systems were available. However, even these limited results, which are summarized by Whitehouse and Hastings (1965), were sufficient for the conclusion to be drawn that $P_1 c(a \to +) \gg P_2 c(a \to +)$ and that $P_2 c(b \to +) \gg P_1 c(b \to +)$. (Analysis of the R classes from this table is discussed in 3.7, and of the t tetrads in 3.9.) Results of outstanding value for their completeness and informativeness have been obtained more recently for *Sordaria fimicola* (Fields and Olive, 1967) and *Saccharomyces cerevisiae* (Fogel and Hurst, 1967). Besides flank markers, an additional (outside) marker at the distal end also was used in

TABLE 5. Classification of 12 Types of Tetrads Containing Wild-Type Spores and Formed in the Cross $Ma + N \times m + bn$

Event in the gene	Configuration of the flanks			
	P_1	P_2	R_1	R_2
$c\ (a \to +)$	$P_1 c\ (a \to +)$	$P_2 c\ (a \to +)$	$R_1 c\ (a \to +)$	$R_2 c\ (a \to +)$
$c\ (b \to +)$	$P_1 c\ (b \to +)$	$P_2 c\ (b \to +)$	$R_1 c\ (b \to +)$	$R_2 c\ (b \to +)$
t (reciprocal recombination)	tP_1	tP_2	tR_1	tR_2

Note. The letters P and R, as usual, denote the configurations of the flank markers (Fig. 26) in the "wild-type" chromatid ($a^+ b^+$).

both systems. In the first system the proximal flank marker was the centromere situated 8.8 map units from the 6A region, while the distal marker was the morphological mutant R 155 (0.25 map unit from it) and the outside marker also was a morphological mutant, R 83 (1.2 map units more distally than R 155). The behavior of the outside marker will be considered later (3.8). From ten crosses 369 asci containing wild-type spores were obtained, and these could be classified as follows: 104 $P_1 c(a \to +)$, 34 $P_2 c(b \to +)$, 50 $R_1 c(a \to +)$, 17 $R_2 c(b \to +)$, 17 $R_1 c(b \to +)$, 4 $P_2 c(a \to +)$, 142 tR_1, and 1 tP_2. The results obtained for the *hi-1* system (histidine synthesis) are given in Table 6. At present we are interested only in the Pc tetrads. The Rc and t tetrads are analyzed in 3.8 and 3.9 respectively. What conclusions can be drawn from these results? First, that the conversion hypothesis of negative interference is very well supported by these results. The P_1 class appears

TABLE 6. Classification of Recombinant Asci of *Saccharomyces cerevisiae* Containing Wild-Type Spores Obtained by Crossing Different Alleles of the *hi-1* Gene (from results obtained by Fogel and Hurst, 1967)

Event in the gene	Number of asci with corresponding configurations of flank markers in the wild-type chromatid			
	P_1	P_2	R_1	R_2
$c (a \to +)$	223	7	138	84
	117	0	82	35
	93	0	43	25
Total (in %)	40	0.7	24	13
$c (b \to +)$	0	23	49	3
	0	16	15	3
	0	10	13	1
Total (in %)	0.0	5	7	0.7
t (reciprocal recombination)	5	0	70	1
	1	0	21	0
	0	0	3	0
Total (in %)	0.5	0.0	8	0.1

Crossing scheme:

```
      +         a         +        ar₆
 thr₃      +         b         +
       I         II        III
              Intervals
```

Flank markers: thr_3 (threonine synthesis) lying 2.5 map units proximally, and ar_6 (arginine) 10.0 map units distally. Numbers in first line of each class apply to the diploid Z 1958 (*hi-1-7* × *hi-1-1*), in the second line Z 2367 (*hi-1-315* × *hi-1-1*), and in the third line to Z 2433 (*hi-1-315* × *hi-1-204*).

Fig. 45. Polaron model of intragenic recombination (after Lissouba et al., 1962). Black squares mark the boundaries of the polaron. Remainder of legend as in Fig. 42.

only as a result of conversions in the proximal site a, and the P_2 class predominantly as a result of conversions in the distal site b. The second conclusion follows directly from the first, and it is that the asymmetry of the flanks observed on statistical analysis is in fact the result of polarity of the conversions in the gene. Consequently the problem is to explain the polarity of intragenic recombination. Can this be done within the framework of the Pritchard–Freese–Murray hypothesis? The answer is evidently no.

According to this hypothesis the frequency of miscopying (conversion) must be low at both ends of the gene and reach its maximum at the middle (Fig. 43), so that the probability of this event depends on distance: the closer the site to the boundary of the fixed region of effective pairing, the lower the probability that it will be copied too often. The types of polarity in the gene predicted by this hypothesis are thus opposite to those found experimentally. In addition, the polarity and asymmetry recorded must depend on which of the crossed markers lies nearer to the center of the gene, and this also is not observed.

An attempt to overcome these difficulties was made by French geneticists (Lissouba, 1960; Lissouba et al., 1962), who totally rejected the multiple switch hypothesis. They introduced the term polaron to denote a fixed segment of the chromosome in which a one-way gradient (polarization) of recombination events (copy-switching, conversions) takes place and in which, as they postulated, recombination is entirely nonreciprocal. Polarity of recombinations within this region is connected with the polar character of DNA synthesis (Fig. 45). Within the polaron there are not many switches, but only one,* and the probability of switching rises from 0 at the beginning

---

*Total rejection of frequent switches within the polaron led to contradictions which will not be considered here. For this reason, Stadler (Stadler, 1963; Stadler and Towe, 1963) proposed a modified varient of the polaron hypothesis. This hypothesis is the last desperate attempt to explain conversion on the basis of copying with exchange of templates. The structural features of Stadler's hypothesis will be considered in 4.7. From the standpoint of formal genetics it is a compromise between Freese's hypothesis and the polaron hypothesis. Accepting the concept of the polaron, Stadler postulated that the number of switches which can take place within it is not one (as in the original polaron hypothesis), and not many, as Freese postulated, but a maximum of two. Sometimes there is one switch and the miscopying continues as far as the end of the polaron, while at other times the miscopying is corrected by a second switch following immediately after the first. The relative frequency of the single and double switches can vary in different systems. Similar ideas have also been developed by Bernstein (1962, 1964).

of the polaron to 1 at its end. Switching leads to miscopying (and, consequently, to 3 : 1 segregation) of the whole terminal part of the polaron from the switching point. Reciprocal recombination (crossing-over) takes place only at special links between the polarons where reciprocal exchanges of the replicas are possible. The polaron hypothesis thus showed a tendency which was already apparent in Freese's hypothesis — rejection of a single mechanism of genetic recombination and the acceptance of two mechanisms (reciprocal and nonreciprocal copy-switching). A similar idea had been expressed previously by Taylor (1958) on the basis of chromosome models developed by Freese (1958) and himself, in which the chromosome was represented as a sequence of DNA molecules of limited size, linked by covalent bonds through protein. The polaron could be identified with the DNA molecule and the special links with protein bonds. In the case of bipolar genes, of course, this idea has an absurd outcome: a gene consisting of two polarons, i.e., divided by a special bond. Moreover, as will be shown in 4.2, the special bond may evidently be a special nucleotide sequence. However, this gives rise to another difficulty: in the bipolar gene DNA replication must start at the special bond and proceed in opposite directions toward the ends of the gene. At the same time, however, the polaron hypothesis (in Stadler's modification) gave the most adequate interpretation of the genetic data.

Despite the fact that the molecular mechanism of conversion in the polaron model was wrong, development of the formal-genetic aspects of this model have played a considerable role in the subsequent development of genetic thinking (Chapter 5). The discovery of the polarity of recombination events and creation of the concept of the polaron thus marked the third important step in the development of the theory of intragenic recombination after the discovery of high negative interference and the hypothesis of the discontinuity of chromosome recombination.

3.7. The Relationship between Conversion and Crossing-over

The conversion hypothesis of high negative interference explains the formation of R classes is the standard flank system by invoking the further assumption that a connection exists between conversion and crossing-over. According to this hypothesis, wild-type recombinants, with the R_1 configuration of their flank markers, appear not as the result of crossing-over between alleles, but as the result of more complex events: conversions at a particular site, accompanied by reciprocal exchanges (crossing-over) on either side of the locus (Fig. 39). Tetrad analysis confirms this hypothesis (Table 6). It is apparent, to begin with, that by contrast with the normal (nonconvertant) ascus, in which any chromatid takes part in exchanges with equal probability, in convertant asci chromatids in which conversion has taken place participate in the overwhelming majority of cases in the exchanges. Crossing-over at flank markers not affecting the "prototrophic" chromatid takes place at a frequency of below 10% (Kitani et al., 1962; Stadler and Towe, 1963; Fields and Olive, 1967; Fogel and Hurst, 1967).

The authors of the conversion hypothesis attributed the connection between conversion and crossing-over to the presence of limited regions of effective interaction between homologous chromosomes, within which the recombination events (either conversion or crossing-over) take place. As an alternative they suggested (Mitchell, 1955, 1957) that favorable conditions for conversion may arise close to the site of crossing-over.

Mitchell's hypothesis was based on the acceptance of two recombination mechanisms. She assumes that conversion may also take place without crossing-over. Freese (3.5) showed that conversion and crossing-over can be explained on the basis of a single mechanism. Within the framework of this "conversion without crossing-over" hypothesis there is a purely formal term to describe the P configuration of the flanks in the convertants, which is the consequence of the double exchange. The results given in Table 6 can thus be analyzed in two ways depending on which hypothesis the analysis is based (Fogel and Hurst, 1967; Stadler et al., 1970).

The Hypothesis of Multiple Exchanges. This hypothesis assumes that all recombinant asci arise as a result of obligatory exchange (for example, switching of the replica* between mutant sites). These exchanges are often accompanied by conversions of the alleles (for example, by excess copying). Comparison of the frequency of appearance of the various types of recombinant asci (Table 6) with the expected frequency of their appearance calculated on the basis of standard genetic lengths of the corresponding intervals (interval *I*: $thr_3-hi\text{-}1$; interval *II*: $hi\text{-}1\text{-}a-hi\text{-}1\text{-}b$; interval *III*: $hi\text{-}1-ar_6$), leads to the following conclusions:

1. Strong positive interference is observed in the distal (*III*) region of the t asci (1 P_2 tetrad instead of the expected 20) and it is absent in the proximal region.
2. In $c(a\rightarrow+)$ tetrads a 15-fold excess of exchanges is found in the proximal interval (*I*), interference is absent in region *III*, and a sharp excess of double exchanges is seen at the flanks. Crossing-over nearly always affects the convertant chromatid.
3. In $c(b\rightarrow+)$ asci, on the other hand, interference is absent in region *I*, but there is a twofold excess of exchanges in the distal region and an increased frequency of double exchanges at the flanks. The prototrophic chromatid is involved in 100% of the exchanges.

The Conversion Hypothesis. This can be called the hypothesis of minimal exchanges for it is based on the assumption that crossing-over takes place only when absolutely necessary (in the formation of Rc tetrads and of

---
*For another possible mechanism, see 5.2.

t tetrads). Analysis based on this hypothesis leads to the following conclusions:

1. Positive interference is observed in the distal region but is absent in the proximal region in t tetrads.
2. A sevenfold increase in the number of exchanges in region I and absence of interference in region III are found in $c(a \rightarrow +)$ asci.
3. In $c(b \rightarrow +)$ asci a threefold excess of exchanges is found in region III and interference is absent in interval I.

The results of analysis are qualitatively similar in each case. Only quantitative differences are found between the models. In either case it is perfectly clear that the formation of prototrophs in interallelic crosses is accompanied by exchanges in the immediate vicinity of the investigated locus, leading to recombination of the flank markers. The results show clearly that the connection between conversion and crossing-over is not accidental in character: conversion of the proximal allele is connected with crossing-over in the proximal region, while conversion of the distal allele is connected with crossing-over in the distal region. The essential fact is that conversion does not interfere with crossing-over on the opposite side of the gene (i.e., conversions in distal sites do not interfere with exchanges in the proximal region, and vice versa). Positive interference, characteristic of crossing-over, is found only in t tetrads. For this reason it is easiest to interpret the origin of t tetrads on the basis of intragenic (interallelic) crossing-over (3.9).

In conclusion, let us imagine that recombination between *hi-1* alleles was studied in random samples of spores. Instead of 12 classes (Table 6) only 4 would have been obtained: $P_1 = 40.5\%$, $P_2 = 5.7\%$, $R_1 = 39\%$, and $R_2 = 13.8\%$, and these could have been analyzed in the same way as any other results of this type (2.8). Clearly in this case the analysis would have been far from complete. Correlation of events within each tetrad or octad gives much additional information.

3.8. One or Two Mechanisms of Genetic Recombination?

Despite the sharp difference between intragenic and intergenic recombination, it was originally thought that both types of genetic recombination could be explained by the single copy-choice mechanism (3.5). It will be shown in Chapter 5 that this is also possible within the framework of the breakage and reunion mechanism. Tetrad analysis of intragenic recombination (3.3–3.6) leads to the conclusion that the molecular mechanisms of crossing-over and conversion are fundamentally different. From the time that conversion was discovered and, in particular, after it had become clear that conversion is the principal (3.4) method of recombination within the gene, attempts have repeatedly been made to define the phenomena of conversion and crossing-over on an other than purely formal basis.

Let us first consider the question of the genetic relationship between conversion and crossing-over. If a single mechanism lies at the basis of both events, conversion must affect crossing-over in neighboring areas in exactly the same way as crossing-over in one area of the chromosome affects subsequent exchanges in neighboring areas, i.e., it must interfere with them. Within the framework of the single hypothesis, any conversion event, whether accompanied or not by recombination of the flanks, must interfere with crossing-over. According to the hypothesis of the double mechanism, only conversion accompanied by recombination of the flanks can interfere with crossing-over (because P configurations of the flanks are formed without crossing-over). To choose between these possibilities, the flank system must be supplemented by an outside marker and the effect of different types of conversions (accompanied and not accompanied by recombination of the flanks) on the frequency of crossing-over must be studied in the interval between them and the nearest flank marker. Observations of this type were first made by Stadler (1959a) in the course of a statistical analysis of intragenic recombination within the cys gene (cysteine synthesis) of $N.$ $crassa$ (Table 7).

Stadler selected cys^+ recombinants obtained in the progeny from cys_t $\times cys_c$ crosses and determined the relative frequencies of appearance of cys^+ with different combinations of flank and outside (ad) markers. He found that $rf_{ylo\text{-}ad}$ in recombinants with the P configuration of flank markers is 13.3%, but only 3.9% in recombinants with the R configuration of the flanks (Table 7). If cys^+ recombinants appeared as a result of conversion, this fact could indicate that the simple event of conversion (without recombination of the flanks) does not interfere with neighboring crossing-over, whereas conversion linked with crossing-over in this site interferes with subsequent exchanges (the result predicted by the two-mechanism hypothesis).

The most demonstrative results were obtained with the $hi\text{-}1$ (histidine synthesis) flank system in yeast (Fogel and Hurst, 1967). The control frequency of exchanges in the outside interval $ar_6\text{-}tr_2$ (where tr_2 is the gene

TABLE 7. Relationship between Frequency of Recombinations Between Markers ad and $yloA$ and Events in the cys Gene of $Neurospora\ crassa$ (Stadler, 1959a)

Configuration of flanks in cys^+ recombinants	$rf \cdot 10^3$			
	Order of markers			
	ad	$yloA$	$cys_c\ cys_t$	lys
P	133	—	—	
R	39	—	—	
Control	138	183	2—3	

Note. ad — adenine synthesis; cys_t and cys_c — alleles of the cys (cysteine synthesis) gene; $yloA$ (yellow conidia) and lys (lysine synthesis) — flank markers.

TABLE 8. Crossing-over in the Outside Interval ar_6-tr_2 in 1081 Asci Containing Spores Recombinant with Respect to *hi-1* Alleles and in 3030 Asci Not Containing Spores Recombinant with Respect to *hi-1* Alleles (control) (after Fogel and Hurst, 1967)

Types of asci	Frequency of exchanges in interval ar_6-tr_2 (IV) in asci	
	With P configuration of flanks	With R configuration of flanks
t	0.250	0.260
c $(a \to +)$	0.400	0.240
c $(b \to +)$	0.450	0.270
Control	0.426	0.186

Scheme of cross

```
  +        hi-1-1       +         ar6       tr2
 thr3  .    +         hi-1-2      +          +
    I          II           III         IV
                    Intervals
```

controlling tryptophan synthesis) is 42.6% (Table 8). Their frequency falls to 18.6% in asci with the P configuration of the flanks, but nonrecombinant with respect to *hi-1* alleles (control) ($C = 0.44$). In recombinant asci with respect to *hi-1* alleles the frequency of exchanges in this interval depends on what events take place in the *hi-1* locus (Table 8).

The results afford convincing evidence that conversion events unaccompanied by recombinations of flank markers do not interfere with crossing-over in neighboring areas, whereas conversions accompanied by recombination of flanks and reciprocal recombination events between alleles (t tetrads) do interfere ($C = 0.61$). Although the facts described above are of considerable value, they nevertheless are not absolute confirmation of the dual-mechanism hypothesis. The single mechanism hypothesis can be formally modified so as to account for the absence of interference (Stadler, 1959a; Ephrussi-Taylor, 1961).*

More weighty evidence in support of the dual nature of the mechanism of genetic recombination was obtained in experiments to study the effects of various factors on conversion and crossing-over. The first experiment of this type was due to Mitchell (1957). She investigated the effect of heat shock on the frequency of interallelic recombination and crossing-over in *N. crassa*. She exposed perithecia or conidia to a high temperature (60°C) for 60 and 240 sec. As a result the frequency of recombinations in the pyridoxin locus

---

*This statement is valid only with respect to the copy-choice hypothesis. Within the framework of the breakage–reunion mechanism, such modification is impossible (5.2).

(alleles *pdx* and *pdxp*) was reduced by 3–5 times, whereas there was no appreciable change in $rf_{pyr\text{-}co}$ (5.7% in the control, 5.3% in the experiment). Similar results were obtained with *Ascobolus immersus* (Lissouba, 1960). Irradiation of *Chlamydomonas reinhardi* with nonlethal doses of γ-rays at the pachytene stage leads to an increase in the frequency of both intragenic recombination (in the *acetate-14* locus) and in crossing-over. On irradiation of zygospores in the S phase, the frequency of intragenic recombination likewise increases, but the frequency of crossing-over decreases (4.5). The frequency of mitotic recombinations rises sharply as the result of ultraviolet irradiation. The effectiveness of induction of conversion is ten times higher than that of crossing-over (4.10). Moreover, some agents (fluorodeoxyuridine, for example) induce only mitotic crossing-over and not interallelic recombination (Beccari et al., 1967). Mitotic conversion of heterozygous diploid cells to homozygous can be the result of crossing-over between gene and centromere (1.6) or of conversion in the heterozygous locus. The frequency of homozygotization in yeast is increased by the action of ethylmethanesulfonate from 1×10^{-4} to 5×10^{-2} (Roman, 1967). In less than 5% of cases this takes place by mitotic crossing-over, and in the rest by conversion. At the same time, homozygotization in most (more than 75%) cases resulting from the action of ultraviolet (Roman, 1967) and x-rays (M. S. Esposito, 1968), is the result of mitotic crossing-over. Nitrogenous bases have a well-marked allele-specific effect on the frequency of conversions in *Sordaria fimicola* and do not change the frequency of intergenic recombinations (Kitani and Olive, 1970).

Another argument in support of the hypothesis that the mechanisms of conversion and crossing-over differ is the following fact. Conversion in meiosis is accompanied in about half of the cases by recombination of flank markers (2.8; 3.8). This correlation is not present in mitosis: both spontaneous and induced convertants appear principally with the P configurations (Fig. 26) of the flank markers (Roman, 1956; Roman and Jacob, 1958; Holliday, 1968). This fact makes an important contribution to the understanding of the nature of the processes leading to conversion and crossing-over (5.6).

Much progress has recently been made in the study of the control of recombination by the gene. Work in this direction has been undertaken now for a considerable time (Detlefsen and Roberts, 1921), but the results obtained in this connection with *Drosophila* have not ruled out the possibility of other interpretations (the Schultz–Redfield effect; see the surveys by: Suzuki, 1963a; Lucchesi and Suzuki, 1968; Valentin, 1972); only very recently has work been published (Chinnici, 1971; Kidwell, 1972) in which the presence of polygenic control over the frequency of recombination in *Drosophila* was demonstrated by eliminating the influence of structural changes in the chromosomes by the microscopic investigation of polytene chromosomes. It has also been shown that differences in the frequencies of crossing-over in males and females of the species *Drosophila ananassae* are due to the presence of an S gene — a dominant suppressor of crossing-over — in the former (Hinton, 1970). Clear and unequivocal results were obtained with fungi. In *N. crassa* (by contrast with *Aspergillus nidulans*) variations in the frequencies of both crossing-over (Frost, 1961; Lavigne and Frost, 1964) and of conversion (Catcheside et al., 1964) are so considerable in different crosses that the map units are virtually meaningless. It is suggested that this is because of heterogeneity of the strains, which are of mixed genetic origin. During inbreeding* the scatter was in fact considerably reduced, and in addition, a marked increase in the frequency of crossing-over and a decrease in the degree of chromosomal interference were observed (Stadler and Towe, 1962; Cameron et al., 1966; Rifaat, 1969). Similar observations have been made on *Schizophyllum* (Simchen and Connolly, 1968).

It was important to discover whether genetically controlled changes in the frequency of crossing-over are accompanied by corresponding changes in the frequencies of conversions, and vice versa. A comparative study of this type was first undertaken by Stadler (1959b). He isolated strains of *N. crassa* in which the frequency of crossing-over between the flank markers *un* and *ylo* differed by twice or three times. However, the frequency of formation of cys^+ prototrophs in interallelic crosses ($cys_c \times cys_t$) remained unchanged. Similar investigations are at present being undertaken on *N. crassa* by Catcheside's group [see also investigations on yeasts (Simchen et al., 1971), on *Schizophyllum* (Schaap, 1971), and on *Ascobolus* (Rossignol, 1972, cited by Leblon, 1972b)]. It was initially postulated that differences in *rf* observed for the same pairs of alleles in the loci *his-1, his-3,* and *his-5* during crossing of independently isolated strains are due to the fact that these strains differ from each other in the gene influencing genetic recombination (the *rec* gene). The dominant allele of this gene (rec^+) suppresses recombination in the *his-1* locus by ten times, in the *his-3* locus by five times, and in the *his-5* locus by half. No effect of rec^+ on crossing-over could be found, but the distribution of flanks in the his^+-3 recombinants was considerably altered (Catcheside et al., 1964).

A careful study of this system showed that not one, but several *rec* genes can be identified in it. A gene (*rec-1*) specifically affecting recombination in the *his-1* gene was discovered first (Jessop and Catcheside, 1965). The presence of the rec^+ allele reduced the scale of the *his-1* gene map by more than 20 times. At the same time (and this is particularly important in this context), intergenic recombination is not affected by this gene. The attempt to study the effect of *rec-1* on recombination in other genes led to the dis-

---

*Crosses of close relatives, leading to homozygotization of the strain.

covery of the *rec-x* and *rec-w* genes (Catcheside, 1968; Catcheside and Austin, 1969). The dominant *rec-x*$^+$ allele reduces recombination by five-sixths in the *his-2* gene. Also, *rec-w* acts on the *his-3* locus. Genes *rec-4* and *rec-6*, which also act on this locus, were isolated independently (Iha, 1967, 1969). It was established later that *rec-x* is identical with the *rec-3* gene controlling recombination in the *am-1* locus (Catcheside and Austin, 1971). Each *rec* gene evidently acts on several different genes, scattered all over the chromosome, at the same time.

Smith (1965, 1966) identified an *rec-2* gene controlling crossing-over in the *pyr-3–leu-2* region (the right-hand arm of the fourth linkage group). The *rec*$^+$ allele of this gene reduces $rf_{pyr\text{-}leu}$ from 21.48% to 10.38% and $rf_{pyr\text{-}his\text{-}5}$ from 14% to 1% in rec_2^+/rec_2 crosses. Although *rf* between the *his-5* alleles (K 512 and K 553) varied in these crosses from 5.28×10^{-5} to 12.7×10^{-5}, no correlation could be found with the *rec-2* genotype of the parents. The impression was obtained that the *rec-2* gene does not affect intragenic recombination in *his-5*. During the study of the effect of genes controlling the repair systems of the cell on genetic recombination (4.9) Holliday (1966b) isolated several mutants of *Ustilago* sensitive to ultraviolet light, and studied both spontaneous and ultraviolet-induced crossing-over and intragenic recombination in them. In one of them (*uvs-2*) spontaneous crossing-over was found to be suppressed, while gene conversion took place at the normal frequency. A similar mutant has been isolated from yeast (*rec-4*; Rodarte-Ramon, 1972). A mutant of *E. coli* affecting intragenic recombination only, and not intergenic, has also been isolated (Zlotnikov and Khmelinskii, 1973) (see also pp. 141 and 224).

Recently (Angel et al., 1970) more complex relationships between intra- and intergenic recombination have been discovered. The degree of increase in *rf* in the *his-3* locus in the absence of the *rec-w*$^+$ allele has been found to be determined by the state of the *cog* locus, situated 1.3 map units distally to *his-3*. In *cog/cog* crosses *rf* is increased by not more than five times, but in the presence of the *cog*$^+$-allele it is increased by 30 times. At the same time the frequency of crossing-over increases in the region between *his-3* and the distal marker *ad-3* (from 1.7 to 4.9 map units). This is the first case in which one system controls both intragenic and intergenic recombination. The *rec-w*$^+$ allele is dominant and epistatic with respect to *cog*$^+$, and in the presence of the former, no allelic differences relative to *cog* are exhibited. The hypothetical scheme of regulation of recombination in this region of the chromosome is illustrated in Fig. 46.

The facts described in this section are thus convincing evidence of differences between the mechanisms of conversion and crossing-over as was suggested by Winge (1955) and Mitchell (1955), and they thus give reliable support to the conversion hypothesis of high negative interference.

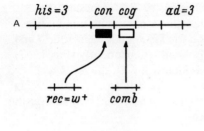
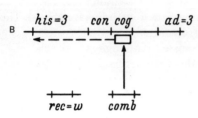

Fig. 46. Diagram showing regulation of recombination in the *his-3* locus of *Neurospora crassa* (after Angel et al., 1970). A) Recombination suppressed, B) recombination allowed (symbolized by interrupted arrow); *rec-w*) regulator of recombination producing the repressor (black rectangle); comb) locus controlling synthesis of recombinase (white rectangle); *cog* and *con*) areas of specific sorption of recombinase and of repressor.

3.9. Origin of Tetrads Containing Reciprocal Products of Recombination between Alleles

In the early stages of the study of intragenic recombination it was suggested that two mechanisms of recombination may operate within the gene: crossing-over, leading to the appearance of a R_1 configuration of the flanks, and conversion, giving the P configuration (Murray, 1960). Tetrad analysis of flank systems showed that the formation of the R_1 class in most cases also is accompanied by conversions. This is evidence against the suggestion mentioned above. However, its possibility cannot be completely ruled out because of the existence of tetrads containing reciprocal products of recombination (t tetrads), and, in particular, because they appear mainly in the R_1 class (the tR_1 class in Table 6) and they interfere with crossing-over (3.8). The idea that t tetrads appear as the result of intragenic crossing-over is now supported by most geneticists (Emerson, 1966b; Fogel and Hurst, 1967; Whitehouse, 1967a; Holliday, 1968). Earlier (Whitehouse and Hastings, 1965), however, another possibility had been discussed and, in the writer's view (5.6), it has not lost its importance with the discovery of the facts just mentioned, namely, that t tetrads arise through reciprocal (in Winkler's terminology, digenic) conversions. As the basis for their discussions Whitehouse and Hastings made certain theoretical assumptions (5.3). However, similar views can be arrived at in another way.

As was mentioned above, during sac formation in some ascomycetes meiosis and the formation of four haploid spores is followed by mitotic (postmeiotic) division, as the result of which each ascus contains eight spores

TABLE 9. Number of Different Types of Aberrant Asci Obtained from g_1 × + Crosses (from data obtained by Kitani et al., 1962)

Segregation in the octad		Number of octads
Number of + spores	Number of g_1 spores	
6	2	98
2	6	13
5	3	108
3	5	20
4	4	9 *

*Octads with a normal 1+ : 1− ratio, but with an aberrant (Fig. 49) arrangement of the spores (in 150 aberrant asci from the total sample).

(an octad). Since recombination takes place at the four-chromatid level, it was evident that each of the four spores is genetically pure, and that the sister spores formed as the result of postmeiotic division must have completely identical genotypes. Accordingly, in early investigations with biochemical mutants of *N. crassa* only one of the pairs of sister spores was usually analyzed. As soon as the morphological characters began to be studied, it immediately became clear that this was not feasible. Olive (1959) found that if a g_1 (allele of the gray gene controlling ascospore color in *Sordaria fimicola*) mutant was crossed with the wild type, 5+ : 3− and 3+ : 5− asci appear just as frequently as 6+ : 2− or 2+ : 6− asci (Table 9; Fig. 47).

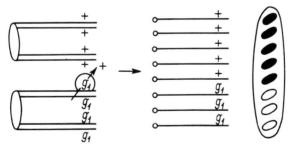

Fig. 47. Scheme of formation of an ascus with a nonidentical pair of sister spores as a result of semichromatid conversion (in crosses between the g_1 mutant of *Sordaria fimicola* with the wild type). Each line on the left-hand side of the figure represents a half-chromatid, and on the right a chromosome. The semicircle with an arrow symbolizes semichromatid conversion.

The 5+ : 3− (or 3+ : 5−) segregation indicated that one of the spores formed as the result of meiosis was heterozygous and that, consequently, conversion affected only half of the chromatid (semichromatid conversion). Initially attempts were made to explain the 5 : 3 segregation by extra-mitosis or by aneuploidy. Cases of the appearance of aberrant asci of this type, caused by heteroploidy and other cytological anomalies, were found experimentally (Case and Giles, 1964), although as was expected, markers linked with the allele segregated in this manner also segregated aberrantly. Fortunately the g_1 locus occurs in a linkage group which has already been studied in considerable detail, together with the other genes determining morphological characters. Since flank markers in most asci of the 5 : 3 type segregated normally, this was evidence in support of semichromatid conversion.

As a result of much work undertaken on *N. crassa* and *Ascobolus immersus* in order to detect postmeiotic segregation not only was its existence convincingly demonstrated, but it was also shown to possess allele specificity. In two-point crosses semichromatid conversion of one allele leads to the formation of octads with a spore ratio of 7− : 1+ (Fig. 48). Octads of this type are found only in crosses involving certain mutants of *A. immersus* (for example, *277* and *1216* from series *46* and *60* from series *19*). Further evidence of allele specificity is given by the fact that mutant *65*, a homo-allele of *60*, does not show postmeiotic segregation. The relative frequencies of formation of the different types of aberrant asci also are allele-specific. For instance, the h_1 mutant (allelic with g_1) does not in general form 6 : 2 octads, but only 5 : 3 and 4 : 4 asci with an aberrant spore distribution (Kitani and Olive, 1967). Postmeiotic segregation is also found in *N. crassa*. Strickland (1961) crossed an inositol-dependent mutant (*inos-37,401*) of *N. crassa* with the prototrophic strain and found one ascus in the progeny with the 5+ : 3− segregation and two with the 2+ : 6− segregation. Allele *64* of the *cys* gene (cysteine synthesis) also shows postmeiotic segregation (5 asci

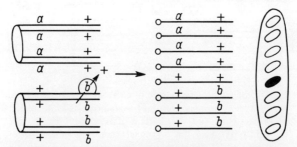

Fig. 48. Diagram of formation of a recombinant ascus in a two-point ($a \times b$) cross as the result of semichromatid conversion of one of the alleles (*b*). Legend as in Fig. 47.

TABLE 10. Number of Mutants of Different Origin with the Same Type of Segregation (after Leblon, 1972a)

Origin	Asci with postmeiotic segregation			
	Rare (1-5%) and excess of conversion		Frequent (up to 90%) and excess of conversion	
Group	K+ A	K− B	K+ C	K− D
ICR_{170}	0	25(3)*	0	0
NTG	0	0	18(6)	0
EMS	4	2	13(8)	0
Spontaneous	2	2(1)	2(2)	2

*Number of mutants with $fk_{a \to +} = fk_{+ \to a}$.

of the 6+ : 2 cys type and 2 asci of the 5+ : 3 cys type among 1637 asci). Semichromatid conversion was not found for other alleles of the cys gene (Stadler and Towe, 1963). All the w mutants of Ascobolus immersus (Pasadena) so far studied (Emerson and Yu-Sun, 1967) exhibited postmeiotic segregation. It is also a characteristic feature of all mutants in the 6A region of Sordaria.

Leblon (1972a,b) has obtained evidence that the relative frequencies of appearance of different types of asci in single-site crosses may depend on the nature of the mutation. Mutants for ascospore color in genes b_1 and b_2 of Ascobolus, induced by the acridine dye ICR_{170} (which gives rise to insertions or deletions of nucleotides), differ sharply from mutants induced by nitrosoguanidine (NTG), producing interchanges of bases, or by ethyl methanesulfonate (EMS), giving large deletions (Table 10).

By contrast with these results, all mutations in segment 29 of locus 14 of Podospora anserina induced both by ICR_{170} and by NTG gave little postmeiotic segregation (Touré, 1972a). Furthermore, the work of Bond (cited by Ahmad et al., 1972) showed that six alleles of the buff locus of S. brevicollis gave PMS in single-site crosses with approximately equal frequency. However, in two-site crosses the wild-type spore in 1+ : 7− asci had a P_2 configuration of its flanks in 37 of 39 cases, whereas in 2+ : 6− asci P_1 and P_2 configurations appeared with equal probability (32 and 36 cases). Inversion in the other linkage group led to a sharp decrease in the frequency of PMS (from 28 to 1.7%). Preliminary data on the dependence of the frequency of PMS on the position of the mutation in the gene have also been obtained by Rossignol and Leblon (1972).

In the experiments of Kitani et al. (1962) 9 of the 150 aberrant asci had the normal segregation (4− : 4+) but abnormal alternation of black and white spores (Table 9; Fig. 49). Whereas the other (flank) markers segregated normally, two pairs of sister spores in each of these asci were heterogeneous in color. The impression was obtained that in this case two semichromatid

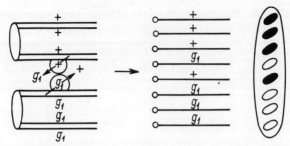

Fig. 49. Ascus with quantitatively normal segregation but with nonidentical spore pairs and the possible mechanism of its formation in crosses between mutant g_1 of *Sordaria fimicola* and the wild type. Legend as in Fig. 47.

conversions took place simultaneously in different directions. Since the two aberrant pairs of spores were constantly found in different halves of the asci, it might be thought that half-chromatids of two homologous (nonsister) chromatids took part in the conversion. However, if the asci with the aberrant 4 : 4 segregation had appeared as the result of random coincidence of the semichromatid conversions, the frequency of their appearance would be the product of the frequencies of the individual semichromatid conversions; since the frequency of appearance of the 5+ : 3− asci is 5×10^{-4}, and the frequency of appearance of the 3+ : 5− asci is 1×10^{-4}, the probability of appearance of asci of the 4 : 4 type is 5×10^{-8}. The frequency observed (7×10^{-5}) is very close to the frequency of formation of asci of the 5 : 3 type. Consequently, if the hypothesis of the mechanism of formation of aberrant 4 : 4 asci is correct, the simultaneous semichromatid conversions must be interconnected. The event observed in that case can be called reciprocal semichromatid conversion. The appearance of asci of the 4 : 4 type with an aberrant arrangement of the spores is thus evidently due to simultaneous semichromatid conversions in two homologous chromatids (Fig. 49). The

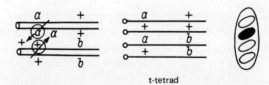

t-tetrad

Fig. 50. Ascus containing reciprocal products of recombination between a and b alleles and the possible mechanism of its formation. Each line on the left-hand side of the Figure represents a chromatid, and on the right a chromosome. Converting alleles are circled. Arrow indicates direction of conversion.

possibility that such a process can take place suggests that t tetrads can be formed by simultaneous chromatid conversions in different directions in homologous chromatids ($a \to +$ and $+ \to a$), as the result of which a "prototrophic" chromatid and a chromatid with the *cis* arrangement of the alleles participating in the crossing appear (Fig. 50). Such an event can be called reciprocal conversion. However, if t tetrads can appear as the result of conversions, the question must be asked: can crossing-over in general take place within the gene? (5.4–5.6).

3.10. Conclusion

Gene conversion was discovered in the 1930s and rediscovered in the 1940s, but the nature of this phenomenon still remains largely conjectural. It is therefore possible to give only a working definition of conversion as a phenomenon revealed by various types of deviation from the Mendelian 1 : 1 segregation in tetrads.

Winkler attempted to use conversion to explain recombination of linked genes, but the genetic significance of this phenomenon only began to be understood in the 1950s through the work of Mitchell, who showed that conversion is evidently a mechanism of intragenic recombination. Further investigations showed that intragenic recombinations are in fact predominantly nonreciprocal, although reciprocal recombination leading to the formation of t tetrads can take place.

If nonreciprocal recombinations are the result of conversion, t tetrads are evidently formed as the result of crossing-over between alleles. On the other hand, the possibility cannot be ruled out that the mechanism of intragenic recombination may be single, on the assumption that t tetrads are formed by reciprocal conversion.

It is previously supposed that the R_1 class in the standard flank system can appear as the result of crossing-over between alleles. However, tetrad analysis of flank systems showed that the R_1 class, like the other classes (R_2 and P classes) is formed by conversion of one of the alleles taking part in the cross. Consequently, conversion is very closely correlated with crossing-over. Events leading to conversion evidently take place at the sites of chiasma formation. Tetrad analysis of intragenic recombination can thus be compared with a powerful ultramicroscope allowing observations to be made at the molecular level on the processes taking place at the sites of chiasma formation.

The answer to the question whether conversion and crossing-over are brought about by different mechanisms or whether they are merely different aspects (manifestations) of the same mechanism is evidently in favor of different mechanisms: sufficient evidence has now been accumulated to show that these phenomena can be differentiated. Investigations into the genetic control of intragenic and intergenic recombinations are particularly interesting and promising (both on their own account and in connection with the solution to this problem).

In the first three chapters we have traced the logic behind the development of genetic thinking which has been based entirely (except for the fundamental fact of chiasma formation in meiosis) on genetic data. Important milestones along this road were the discoveries of polarity and asymmetry of the P classes, indicating discontinuity of the chromosomes in relation to recombination events. At the same time, views regarding the concrete molecular mechanisms of recombination have not yet left the realms of speculation.

Further progress in the study of the mechanism of genetic recombination was achieved through the study of its structural and biochemical aspects. The next chapter deals with these matters.

CHAPTER 4

Experimental Study of the Mechanism of Recombination

4.1. Introduction: The Problem Stated

In the middle of the 1930s most of the evidence for the chiasmatype theory had been obtained. (The final stroke to complete the picture was added by Taylor's experiments; see 4.8.) However, since it was restricted to the level of cytological observations, it could not answer the question of the causes of the breaks in the chromatids which it postulated and which lead to the formation of chiasmata and of crossing-over. Since there were no direct experimental approaches to the study of the mechanism of crossing-over at that time, there was plenty of opportunity for various types of speculation. The hypothesis which enjoyed the greatest popularity was that of Darlington, which claimed to give a complete and final explanation of all the phenomena connected with crossing-over (synapsis, chiasma formation, and interference) on the basis of the assumption that the cell embarks prematurely (compared with mitosis) on meiotic division, as a result of which chromosomes which have not yet divided conjugate because of their inherent "pairing urge." Replication of chromosomes, which begins after cytologically detectable synapsis, creates mechanical stresses in the helically coiled homologues, leading to breakages (1.13). According to Belling, recombination of genes is also caused by replication, but the recombinant chromatids arise, not mechanically, but through late replication of the chromoneme compared with replication of the chromomeres (1.14). Since Belling's hypothesis explained neither synapsis nor interference, the choice in general was decided in support of Darlington's hypothesis, which had the support of most geneticists until the middle of the 1950s.

There was no radical break with the canonical views of the mechanism of recombination (1.19) as the result of a single experiment, but rather a general tendency was induced for research to change to the molecular level as the result of the development of genetics of microorganisms and the decoding of the structure of DNA (Watson and Crick, 1953a). Since Belling's hypothesis (in its modified form; see 1.19) was more in harmony with the general tendency to think in molecular terms, it was soon adopted despite its obvious disadvantages.

Although Darlington was content to consider chromosomes as macromolecular structures possessing specific mechanical properties, the change to the molecular level of description of the recombination process demanded knowledge of the molecular organization of the recombining structures and also of their method of replication appropriate to this new level. It was necessary to discover whether the processes of replication, recombination, and synapsis coincide with each other in the cell cycle, as both hypotheses require, or whether the situation is otherwise, in which case both would have to be rejected. These problems are discussed in the first part of this chapter (4.2-4.7). In the second part of the chapter (4.8-4.10) direct approaches to the study of the mechanism of recombination associated with the use of modern physicochemical and biochemical methods are described. A biochemical concept of crossing-over which is adequate to explain the mechanism of induced recombinations and structural changes in chromosomes is formulated.

4.2. The Structure of Chromosomes

The chromosomes of prokaryotes (organisms without formed nuclei) are double-helical DNA molecules of varied molecular weight (from 5×10^6 to 3×10^9 daltons). Exceptions to this rule are some bacterial viruses whose genetic material consists of RNA or single-stranded DNA (Kleinschmidt, 1967). The chromosome of the prokaryotes is a single replicon: its replication starts at a certain point and continues successively along the entire chromosome until the cycle is complete (Cairns, 1963). The chromosomes of prokaryotes are always either rings or capable of forming rings. Concatenate structures constitute an important part of their reproductive cycle (Skalka, 1971; Watson, 1972.

The DNA content in the chromosomes of eukaryotes is hundreds or thousands of times greater. The DNA is bound with basic proteins (Mirsky, 1971) which determine the complex morphological structure of these chromosomes (Cole, 1962; De, 1968). The relationship between the DNA molecules and the chromosome structure as seen in the light microscope is a relatively new problem, although the number of elementary threads in the chromosome and the method of their organization are questions which were asked soon after chromosomes were discovered. In the light microscope anaphase (Fig. 1) chromosomes often appeared to consist of two, or even four elementary threads (Huskins, 1952; Bayer, 1965; Maquire, 1966b, 1968a). The term elementary thread implies a structure containing all the genetic information of a particular chromosome. The chromosome can be regarded as consisting of several elementary threads (the polynemic structure) or of only one (mononemic structure). In the first case the genetic information in the chromosome is duplicated several times. Determination of the number of elementary threads is particularly important in the case of meiotic chromosomes, for it is in them that changes (mutations and recombinations) of genetic importance arise. Mitotic chromosomes are often polynemic.

The solution of this problem was delayed by the absence of suitable methods of investigating the submicroscopic organization of chromosomes. The introduction of the electron microscope provided the facilities needed for such investigations, but at the same time new difficulties arose in the study of intact chromosomes by this method. To begin with, therefore, progress was limited to the analysis of ultrathin sections of fixed chromosomes and interphase nuclei. It was difficult to recreate the overall picture of a large structure from the analysis of random sections 300-500 Å thick. The photomicrographs obtained usually revealed a shapeless mass of segments of threads and granules of different sizes. The only common element regularly found in these sections was a bundle of threads varying from 30 to 500 Å in diameter. It was concluded from these observations that the chromosome is polynemic in structure, i.e., that it consists of a bundle of coiled elementary threads, the number of which may reach double figures (Ris, 1957). This is a structure reminiscent of a multicore cable. It was postulated at first (Kaufmann, 1957; Ris, 1957) that the diameter of the elementary thread is 100-200 Å and that its structure is similar to that of tobacco mosaic virus (a molecule of nucleic acid surrounded by a protein membrane). Later (Ris and Chandler, 1963) the diameter of the elementary thread was found to be 40 Å. It was postulated that two such threads are firmly bound together, so that threads 100 Å in thickness are most commonly found in the sections.

Although cytologists had thus apparently obtained conclusive evidence of the multistranded structure of the chromosomes, most geneticists were very wary of accepting it. There was indeed a firm conviction that the chromosome cannot be multistranded (Freese, 1958). So far as recombination is concerned the chromatid in fact behaves as a single structure, since two meiotic divisions are usually sufficient for the completion of segregation (postmeiotic segregation indicates further divisibility of the chromatid, which in the single-stranded concept is associated with the double-helical structure of DNA). The absence of delay in the appearance of recessive mutations is also evidence against a multistranded structure of the chromosomes. Fortunately for the supporters of the multistranded concept it was found that mutations induced by chemical mutagens sometimes appear only after several generations (delayed mutations) (Bird and Fahmy, 1953). Other facts such as the existence of "semichromatid breaks" and sharp variations in the content of nuclear DNA in the parental species (Sunderland and McLeish, 1961) were also adduced in support of the multistranded concept. A detailed discussion of these facts is outside the scope of the book, and it will therefore merely be stated that delayed mutations are also found in phages (Kriviskii, 1966), that "semichromatid breaks" are evidently uncompleted chromatid aberrations (Sidorov et al., 1966; Kihlman and Hartley, 1967; see also 4.10), and that variations in the DNA content can be rationally explained by the single-stranded concept (the slave-repeats hypothesis*).

Indirect confirmation of the single-stranded nature of chromosomes is obtained by results showing that the radiosensitivity of plant cells is directly proportional to their DNA content (Sparrow and Evans, 1961). The opposite would be true if the multistranded hypothesis were correct.

Direct cytological proof of the single-stranded nature of chromosomes has recently been obtained. This advance was due to the development of methods of isolation of intact chromosomes from eukaryotes and of their electron-microscopic investigation in thin films (Kleinschmidt et al., 1962; Gall, 1963a). This method corresponds to the squash technique in ordinary microscopic work. It was shown in this way that chromosomes at all stages of the cell cycle contain a system of threads several centimeters in length (up to one-quarter of the whole chromosome) (Sasaki and Norman, 1966), and about 200-250 Å in diameter (Du Praw, 1965a,b). This mean value is strictly constant for the strands of chromosomes of all species so far investigated. The strands are irregularly arranged in interphase nuclei and meiotic chromosomes, and they show no signs of a helical structure or of a central rod giving off side branches; no half-chromatids likewise have been found (Gall, 1966; Comings and Okada, 1969). The apparent multistranded structure of chromosomes in ultrathin transverse sections is evidently explained by repeated folding of the elementary thread. The absence of coiling proves that differences between the structure of the interphase and anaphase chromo-

---

*This concept was developed originally in order to explain a number of special features observed in the structure and behavior of lampbrush chromosomes (see below). Briefly it can be described as follows: each unit of a genetic function in the chromosome consists of one master-copy. To apply this concept to such well established facts as the existence of not more than two alleles of one gene in the diploid organism, and the fact that *rf* values between sites in neighboring genes may be less than between sites in the genes (Calef, 1957), it is postulated that: 1) any mutations arising in the slave-repeats are quickly corrected back to the original nucleotide sequence of the master-copy; 2) the slave-repeats do not take part in recombinations and they are excluded from the synaptinemal complex (4.6), but they form chromomeres and, subsequently, lampbrush loops. This hypothesis is supported by the fact that chiasmata are never observed in the loops (Callan, 1960). Any changes arising as the result of recombinations or mutations in the master-copy are later reproduced in the slave-repeats (for the mechanism of correction, see 5.6) (Callan and Lloyd, 1960; Callan, 1967; Whitehouse, 1967b).

somes are entirely dependent on the degree of folding (the packing density or condensation) of the elementary thread (Du Praw and Rae, 1966). The chief difficulties in the interpretation of results obtained by the squash technique arise during comparison of the diameter of the elementary thread (not less than 200 Å) with its diameter in ultrathin sections. It has been shown (Wolfe and Grim, 1967) that the technique of preparation of the "squash" preparations is responsible for this discrepancy, for it leads to swelling of the protein membrane of the elementary thread. Its true diameter in the intact nucleus is evidently 100 Å, and no evidence of further subdivision into substructures of a lower order can be found in these threads. Du Praw (1965a,b) confirmed Ris's original suggestion that the elementary thread is formed by a DNA molecule surrounded by a protein membrane: after treatment of chromosomes with trypsin the elementary thread 200 Å in diameter is converted into a thread 30 Å in diameter (which corresponds approximately to the diameter of the DNA molecule).

Convincing evidence in support of the single-stranded concept has been obtained by the investigation of lampbrush chromosomes. These chromosomes, formed from chromosomes of the ordinary type in amphibian oocytes at the diplotene stage, consist of two chromatids each of which forms long side loops (Fig. 51). Gall (1963b) showed that neither ribonuclease nor proteases can destroy the structure of the loops, although they have a high content of protein and RNA. If the protein and RNA are dissolved in concentrated KCl a thin thread, probably a DNA strand, about 30 Å in diameter is left (Miller, 1964). The thickness of the central rod in the region between the chromomeres was found to be 50 Å (see also Miller and Hamkalo, 1972).

Gall (1963b) investigated the kinetics of breakages induced by deoxyribonuclease in the chromosomes of *Triturus cristatus*. He photographed unfixed chromosomes in a special chamber at certain time intervals after immersion in a solution of deoxyribonuclease and counted the number of breaks in the side loops and in the central rod on the photographs. Gall's basic assumption was that the chromosome consists of n complementary DNA strands. Each DNA strand is attacked independently and with the same probability by the enzyme, and they must all be destroyed at approximately identical points for a visible break to be formed. If there is only one double-helical DNA molecule, consisting of two complementary strands, the number of visible breaks must be proportional to the square of the time. Gall ob-

Fig. 51. Morphology of the diplotene chromosomes of *Triturus* (from Gall, 1956).

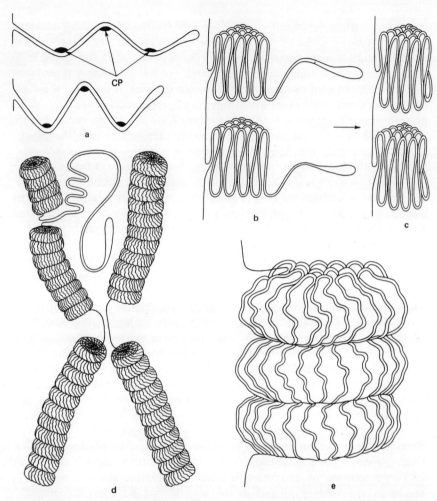

Fig. 52. Model showing folding of DNA in the chromosomes of a eukaryote (from Mosolov, 1968). Each line represents a double-helical DNA molecule. a-c) Successive stages of folding of replicons into parcels (CP, hypothetical contractile protein); d) further condensation of DNA, by compression of parcels of chromomeres; e) general appearance of the metaphase chromosome; one replicon is shown in the decondensed state.

tained values of $n = 2.6 \pm 0.2$ for the loops and $n = 4.8 \pm 0.4$ for the central rod. The true number of subunits is evidently two and four, because there are two factors which tend to make the value of n too high: delay in penetration of the enzyme through the nuclear membrane and delay in discovery of the breaks after they have taken place. The experiments of Taylor (4.3) provided even more conclusive evidence in support of the single-stranded structure of the chromosomes.

There is therefore more than sufficient evidence available at the present time in support of the single-stranded structure of meiotic chromosomes. The chromosome is a single giant DNA molecule or a series of several molecules, each of which is an individual replicon. Although the precise number of replicons in the chromosomes is unknown, the fact that multiple points of initiation of replication exist along the chromosome has been established beyond all doubt (Hsu, 1964; Huberman and Riggs, 1966; Painter, 1966). Their number is conjecturally the same as the number of chromomeres (Pelling, 1968; Lima de Faria and Jaworska, 1972). A possible model of the packing of the giant DNA molecule in the eukaryote chromosome is shown in Fig. 52 (Mosolov, 1968). On packing in the prokaryotes, see Worgel and Burgi (1972).

The discussion on this problem can be summed up on the statement that by no means all geneticists support the single-stranded concept of chromosome structure at the present time (see the surveys by Wolff, 1969; Prescott, 1970; Stubblefield and Wray, 1971). However, the arguments of the supporters of the multiple-stranded model do not seem to the writer to be sufficiently convincing in the light of the facts described above. For our purposes it is convenient to regard the eukaryote chromosome as a single DNA molecule. At the same time, modern hypotheses of crossing-over (Chapter 5) can be applied with equal success to both single-stranded and multistranded models of chromosomes (Holliday, 1968, 1970).

There are now sufficient grounds for postulating two levels of continuous organization of the chromosome in relation to recombination. The first level, which is connected with structural differentiation of the chromosome along its length (centromere, chromomeres, heterochromatin, and euchromatin) will not be examined here in detail (see Chapter 1). For this reason, matters connected with the linear structural differentiation of the chromosome also will not be considered here. I shall merely say that the centromere region has no structural differences revealed by the electron microscope from other parts of the chromosome (Comings and Okada, 1969), that the difference between heterochromatin and euchromatin is determined by the degree of condensation of the elementary thread (Ris, 1957; Perreault et al., 1968), that the latter is richer in G-C (Comings, 1972), and that the chromomeres may be parcels of slave-repeats which do not participate in recombination (see above) (see the survey by Thomas, 1971; see also Lambert, 1972; Judd et al., 1972; and for a model, see Henderson, 1972).

As regards the second level of recombination discontinuity of the chromosome (intragenic: 2.11), attempts have been made to explain it by the presence of special bonds or "linkers" (of protein or other nature) between the DNA molecules (Freese, 1958; Uhl, 1965). For instance, after the discovery of the effect of EDTA on crossing-over in *Chlamydomonas reinhardi* (Russell and Tatum, 1956) it was postulated that the DNA molecules in the chromosomes may be linked together by calcium and/or magnesium bridges. This hypothesis has not been confirmed because it has been shown that ribonuclease has a similar action on crossing-over in the X-chromosome of *Drosophila* (Kaufmann et al., 1957). The effect of both agents on crossing-over is evidently connected with general inhibition of protein synthesis or DNA synthesis (Landner, 1972).

Davies and Lawrence (1967) treated a population of zygospores of *C. reinhardi*, synchronized with respect to their meiotic stage, by various inhibitors of RNA and protein synthesis: chloramphenicol, 5-fluorouracil, 8-azaguanine, and ribonuclease. Cell samples collected at various stages were treated for 20-25 min with these inhibitors in concentrations giving a perceptible effect but without substantially reducing the viability of the cells. Regardless of the stage of meiosis at which the treatment was given, inhibition of protein synthesis was found to lead to a decrease in the recombination frequency. The results obtained in experiments on *C. reinhardi* thus demonstrate conclusively that inhibitors of protein synthesis have a nonspecific effect on crossing-over. If crossing-over took place at the protein "linkers," it would be expected that the effect of inhibitors of protein synthesis on the recombination frequency would be dependent on the stage: for example, it would occur only (or especially) at that stage of meiosis (the pachytene stage) during which the structural proteins essential for the completion of crossing-over are synthesized (Uhl, 1965). The conclusion that no "linkers" of protein or any other nature, are present between the DNA molecules is confirmed by the findings of Solari (1965, 1967), who isolated DNA molecules up to 108 μ in length from sea urchin sperm under conditions leading to protein degradation (detergent and pronase). DNA molecules equal in size to the amount of DNA contained in the largest chromosomes of the anaphase mitotic set of *Drosophila* have been isolated (Kavenoff and Zimm, 1973). The presence of "linkers" has been completely ruled out in bacteriophage chromosomes (Kleinschmidt, 1967). The discontinuity of genetic material postulated above (from the recombination standpoint) can thus be successfully achieved without the role of proteins at the DNA level in precisely the same way as its functional discontinuity (boundaries between genes and replicons) is achieved, for example, with the aid of special nucleotide sequences (Opara-Kubinska et al., 1964). Holliday (1968) calls this hypothetical sequence the recombiner.

4.3. The Character of Replication of DNA and Chromosomes

The semiconservative character of DNA replication postulated by Watson and Crick was first demonstrated in the experiments of Meselson and Stahl (1958) who used the method of DNA centrifugation in a density gradient for this purpose. Both the experiment itself and the method of analysis are now sufficiently well known and adequately described in the literature (Hayes, 1968; Bresler, 1966). The appearance of isotope labeled hybrid DNA has since been frequently observed in experiments on a wide range of organisms starting with bacteriophages (Meselson and Weigle, 1961) and ending with cultures of HeLa cells (Simon, 1961). However, experiments carried out in accordance with this scheme have told us nothing about how the chromosome itself divides because not enough is known about its structure. This problem has been investigated autoradiographically (for a survey, see Taylor, 1963). The results showed that distribution of the label (usually tritium) in mitosis of different organisms conforms to the semiconservative pattern: after one cycle of replication in a medium with tritiated thymidine the original unlabeled chromosome forms two labeled sister chromosomes, each of which segregates in medium without tritiated thymidine into one labeled and one unlabeled chromosome. The relative frequencies of the single and double exchanges observed in these experiments between the sister chromatids can be satisfactorily explained by the hypothesis that each chromatid contains two different subunits (antiparallel DNA chains). This interpretation, however, is not accepted by all geneticists (for a survey, see Heddle, 1969, and for a discussion, Comings, 1971; Gatti and Oliveri, 1973).

The study of the character of DNA and chromosome replication in meiosis was held up for a long time by experimental difficulties. It was only recently that Taylor (1967) succeeded in showing that the distribution of tritium label between chromatids in spermatogenesis in the cricket also is semiconservative in character. At the same time (Sueoka et al., 1967) evidence of the semiconservative character of DNA replication in meiosis was obtained in *Chlamydomonas reinhardi.*

4.4. Identification of the Stage of DNA Synthesis in Meiosis

The classical cytogenetic theories (1.9, 1.13, 1.14) postulated that replication of chromosomes in meiosis takes place in the early pachytene (and even in the diplotene) stage; consequently, the chromosomes have not yet replicated in the leptotene stage. Although many cytologists formerly considered that leptotene chromosomes consist of two chromatids (Huskins and Smith, 1935) this view was rejected for two reasons: the existence of data to the contrary and the simplicity and apparent conclusiveness of Darlington's "prematurity" hypothesis (1.13, 4.1).

It was shown by quantitative cytophotometry of nuclei stained by Feulgen's method and by autoradiography of the nuclei after introduction of labeled DNA precursors (P^{32}, thymidine-H^3) into them that DNA replication in meiosis takes place in a very wide range or organisms during the last premeiotic interphase (the synthetic or S- period), and ends before the onset of the leptotene stage or early in it (Swift, 1950; Taylor, 1953; Henderson,

1964; Moens, 1970). It is also considered that DNA replication takes place during duplication of the chromosomes, i.e., during the formation of two chromatids, for the transition from chromosome to chromatid aberrations coincides with the S-period (Thoday, 1954). Electron-microscopic studies show that leptotene chromosomes are duplex in nature, i.e., they are subdivided into chromatids (Comings and Okada, 1970; Westergaard and Wettstein, 1970). Synthesis of chromosomal proteins also takes place in interphase (Taylor and Taylor, 1953), although doubling of their content does not necessarily coincide with the moment when the cell begins meiosis (Bogdanov et al., 1965).

Recent work has shown that the last cycle of DNA replication in fungi and in some strains of *Chlamydomonas* takes place before physical fusion of the nuclei or gametes (Rossen and Westergaard, 1966; Sueoka et al., 1967). Nevertheless, some investigations showed that incorporation of a radioactive label into chromosomes can be recorded even in the pachytene stage (Sparrow, 1952; Moses and Taylor, 1955), or later still (Riley and Bennet, 1971). This effect was demonstrated most clearly by Wimber and Prensky (1963), who found that up to 2% of the total content of thymidine-H^3 incorporated during the S-period can be detected in the pachytene chromosomes of *Triturus*. Hotta (Hotta et al., 1966; Hotta and Stern, 1971) showed that incorporation of the label into prophase chromosomes is not accounted for by continuation of the S-period until the pachytene stage, but is a different process. To study DNA synthesis, incorporation of P^{32} into microsporocytes of *Lilium* and *Trillium* was investigated. The results showed that 99.5% of DNA in meiotic cells replicates in the S-period. Next follows a pause. In the period between the end of the S-period and the late leptotene stage no significant DNA synthesis is found, but starting from early zygotene DNA synthesis is again observed until its completion in the pachytene stage, after which incorporation of the label into DNA ceases completely. Centrifugation of DNA in a CsCl gradient showed that two events requiring DNA synthesis take place in this period — one in early zygotene, the other in late zygotene and pachytene. In early zygotene DNA of increased density (increased content of guanine and cytosine) is synthesized, whereas in pachytene DNA of the ordinary type is synthesized; however, it is renatured much faster than total DNA (Smith and Stern, 1972).

The functional role of the limited DNA synthesis in meiosis has recently been studied in an elegant piece of research (Ito et al., 1967) on a synchronous population of meiotic *Lilium* cells in culture. DNA synthesis was suppressed by deoxyadenosine. The study of the relationship between DNA synthesis and the development of events in meiosis was facilitated by the fact that cells explanted at the leptotene stage did not enter the zygotene and pachytene stages for several days. The presence of the inhibitor (10^{-2} M) depressed P^{32} incorporation by 91% and somewhat retarded the normal course of meiosis (probably because of inhibition of RNA synthesis by 20%). Inhibition of DNA synthesis at the early zygotene stage led to the arrest of meiosis without induction of any visible anomalies in cells dying after 4-5 days. Inhibition at the middle zygotene stage did not have such dramatic consequences, but fragmentation of the chromosomes in prophase and metaphase was observed in this case. Treatment of cells in the late zygotene–early pachytene stage led to the appearance of "fragility" of the chromosomes and to breakages which did not appear until the anaphase II stage. In mid-pachytene the sensitivity of the cells to deoxyadenosine fell sharply, and treatment in the subsequent stages was completely ineffective. The difference in sensitivity of the stages of meiotic prophase to inhibitors of DNA synthesis is in sharp contrast to the equal sensitivity of all the stages of meiosis (up to diakinesis) to inhibitors of protein synthesis. In the last case the cytological disturbances were found to be of a special type: segregation of the chromosomes is disturbed and is accompanied by changes in their morphology (compare with 4.5).

4.5. At What Stages of Meiosis Do Conversion and Crossing-over Take Place?

Since the work of Janssens it was considered that crossing-over must take place in meiosis. However, there were only two or three experimental studies to confirm this view (Plough, 1917, 1924). It was Plough who first showed the effect of temperature on recombination frequency. He also observed that temperature is a specific agent only for a particular stage of meiosis (cogenesis in *Drosophila*). Plough obtained successive batches of eggs laid by a female exposed to various temperatures and determined the percentage of crossing-over in them. The effects of exposure to temperature were reflected by a waveform curve rising to a maximum. Since each particular batch of eggs corresponded to the action of temperature at a certain stage of development of the primitive cells, after cytological determination of the maturation time of the oocytes it was easy to calculate which is the most sensitive stage. The results showed that both a high (30-32°C) and a low (10-13°C) temperature increase the recombination frequency but only if they act on the early oocyte stages (late leptotene–early diplotene).

However, this work was completely ignored when the protagonists of the copy-choice hypothesis hit upon the fact that DNA synthesis takes place in premeiotic interphase. This fact, together with other considerations (4.6), led them to take a fresh look at the "dogma" that recombination takes place in the prophase of meiosis. Pontecorvo (1958) and Pritchard (1960a) drawing attention to the presence of mitotic crossing-over, advanced the hypothesis that recombination can take place in the interphase nucleus, i.e., during DNA synthesis; consequently, pairing of the chromosomes must take place at this time (for a discussion of this problem, see 4.6). Because of increasing interest in the problem of the mechanism of recombination, experiments have recently been carried out on various objects in an attempt to discover at which stage (or stages) of meiosis genetic recombinations take place (survey: Maquire, 1968b). Grell and Chandley (1965) repeated Plough's experiments, by exposing flies to a temperature of 35°C for 24 h and recording the S-stage by autoradiography of histological sections of the ovaries. These workers observed a maximal increase in the frequency of crossing-over in the region of chromosome *II* adjacent to the centromere in a batch of eggs 9 days after laying. The mean time required for maturation of the labeled oocytes also was found to be 9 days; consequently, crossing-over coincides with the time of DNA synthesis in premeiotic anaphase. However, the intervals of selection used in this investigation were evidently too long to provide adequate resolution, for the interval between the S-stage and prophase in *Drosophila* cannot exceed 1 day in duration. This view is confirmed by results obtained by Henderson (1966), who studied the effect of heat shock on chiasmata in the locust. Meiosis in the locust at 40°C lasts 7 days, and most of this period (5 days) is taken up by the leptotene and zygotene stages, so that the S-stage and pachytene are well separated in time. The locusts were kept at 40°C, and an intraperitoneal injection of labeled thymidine was given 1 min before they were placed in the incubator. Exposure to a constant temperature of 40°C led to a sharp decrease in the frequency of chiasmata; reduction did not begin until the 4th-5th day and reached its maximum on the 6th day. Autoradiographic studies showed that labeled cells reach the diplotene stage 2 days after the reduction in the number of chiasmata begins; consequently, events sensitive to the action of temperature take place, not during the S-period, but at least 2 or 3 days later, i.e., in the zygotene or early pachytene stage. Similar results were obtained on *Goniaea australasiae* by Peacock (1970): the sensitive stage occurs 4 days after the S-phase but 6 days before the pachytene (early pachytene). These results, and also those obtained by Raju and Lu (1973) in *Coprinus lagopus,* showing that temperature induces a maximal increase in the recombination frequency during karyogamy and synapsis, prove beyond all doubt that crossing-over takes place in the prophase of meiosis. It is not yet clear, however, why it causes an increase in *rf* and a decrease in the frequency of chiasmas.

Meanwhile, as the latest work of Grell (1973) and others has shown (see below), the response to temperature may be highly complex: regions located on either side of the centromere, especially in chromosome *3,* react most sharply, while the proximal and distal regions react to a lesser degree, and the inner segments of the chromosomes give hardly any response at all. It has been confirmed for most regions that there is overlapping, at least partial, between the S-phase and the temperature-sensitive stage. Moreover, temperature can influence even earlier stages (see the discussion in Lamb, 1969, 1971; Nevzgladova, 1972; Bayliss and Riley, 1972, 1973).

Lawrence (1965) showed that low-temperature shock increases the recombination frequency in *Chlamydomonas* on the average by 8.5% if the zygospores are exposed to the low temperature 6-7 h after the beginning of development. The S-phase lies between 6 h 30 min and 7 h (Chiu and Hastings, 1973).

Recent work has shown that inhibitors of DNA synthesis have a considerable influence on crossing-over in *Chlamydomonas* (Davies and Lawrence, 1967). Various inhibitors were tested and the results showed that 5-deoxyadenosine inhibits recombination (to 85% of the control level) only if it acts on the 6.5-hour stage after the beginning of zygospore development. Adenine was effective at two stages: the synthetic and the 5.5-hour stage. By contrast with these inhibitors, mitomycin stimulated recombination and its effect reached a maximum with the treatment of 5- and 6-hour zygospores.

In another investigation (Chiu and Hastings, 1973) the maximum of recombination induced by mitomycin C also occurred in the period between 5 and 5.5 h. Fluorodeoxyuridine (FUDR) and nalidaxic acid suppressed recombination by their action on the synthetic stage (6-7 h).

The study of the effect of mono- and bifunctional alkylating compounds on recombination in *Chlamydomonas* showed that diethyl sulfate inhibits recombination at two stages (preleptotene and prophase), whereas nitrogen mustard has no stage-specific action (Davies, 1966).

It was hitherto supposed that ionizing radiation is not so strictly stage-specific as exposure to temperature in inducing changes in recombination frequency in a series of successive ovipositions in *Drosophila* (Whittinghill, 1955). However, as more accurate experiments showed, the action of ionizing radiation is also stage-specific. The work of Zakharov and Inge-Vechtomov (1961) showed that x-rays induce an increase in rf not in all batches of eggs laid, but only in those which correspond to the early oocyte (prophase) and premeiotic, oogonial stages.

The conclusion that the action of ionizing radiation on crossing-over is nonspecific was deduced from experiments in which high doses of irradiation were used, doses which probably disturb the general physiological state of the cells. In addition, it was not always possible to be absolutely sure that the reaction of cells homogeneous as regards their stage of meiosis was being studied. Experiments on *Lilium, Tradescantia,* and *Chlamydomonas* (Mitra, 1958; Lawrence, 1961a,b, 1965, 1967, 1968) showed strictly that low doses of x-rays and γ-rays, although not causing death of the cells, affect the frequency of chiasma formation and crossing-over at two stages of meiosis. The first stage, at which irradiation reduces the frequency of chiasmata and rf, occurs immediately before the beginning of meiosis (preleptotene, or end of the S-phase). The second stage during which irradiation increases the frequency of chiasmata and rf is the late zygotene—early pachytene stage.

All this information was obtained by the study of crossing-over. So far as intragenic recombination is concerned, very little work has been done in this direction. Lawrence (1967, 1968) showed that γ-ray irradiation affects recombination within the gene at two stages also. Whereas recombination in the *acetate-14* gene is increased by 1.8 times in both the preleptotene and pachytene stages, crossing-over is inhibited in the first stage both in the region across the centromere and in the center of linkage group I.

Lewis (1972) found that the effect of an increase in recombination frequency in the *b2* gene under the influence of γ-rays can be obtained only if irradiation is given in the pachytene stage. In *N. crassa* ultraviolet radiation increases the frequency of gene conversion only if given on the 1st or 2nd day after plasmogamy (Hammerl and Klingmuller, 1972). Mitotic recombinations also take place successfully during blocking of DNA synthesis (Holliday, 1971; Unrau and Holliday, 1972).

These facts show conclusively that two stages of meiosis are sensitive to agents of various types: one stage at the end of the S-period (preleptotene) and the other in prophase (zygotene–pachytene). Although recombination in fact takes place in meiosis, the first preparatory stages evidently begin in the S-period or at its end (Mitchell, 1960; Holliday, 1966b; Whitehouse, 1966; Davies and Lawrence, 1967; page 170). The mechanisms whereby these various agents affect crossing-over are discussed in Section 4.10.

The study of meiosis in fungi (4.6) showed that synapsis is induced between maximally condensed chromosomes, but until the end of zygotene, i.e., until the closest degree of conjugation, a small degree of lengthening and decondensation is possible. Lengthening continues in pachytene also, but maximal uncoiling of the chromosome occurs in the stage after pachytene, to which the name "diffusotene" was formerly given (Iyengar, 1939). Next follows sharp contraction (condensation) of the chromosomes and diakinesis with chiasmata. It was postulated (Barry, 1966, 1969) that "diffusotene" is also the stage at which crossing-over takes place, for at this period the chromosomes are in the most favorable state of maximal decondensation for contact formation. However, a clearly defined process of separation of the homologues can sometimes be observed after pachytene and before the beginning of diffusotene (Lu, 1967). It can accordingly be considered that diffusotene is the analog of the diplotene stage in amphibians, at which the chromosomes acquire the lampbrush or "diffuse" appearance on account of the processes of RNA and DNA synthesis, which are intensified during this period of meiosis. This type of behavior of the chromosomes in diplotene (mistakenly called diffusotene) is evidently a characteristic feature, to a greater or lesser degree, of all animals and plants (Henderson, 1964; Moens, 1964). Evidence that crossing-over takes place in fact in pachytene, and in no other stage of meiosis, is given by the presence of the synaptinemal complex (4.6), which is formed in zygotene (after completion of total DNA synthesis: Moens, 1970) and dissociates from bivalents emerging from the pachytene stage.

4.6. Synapsis of Chromosomes

Another aspect of the problem of the place and time of genetic recombinations in the cell cycle is that of when conjugation between homologous chromosomes takes place. If recombinations coincide in time with DNA replication, as the copy-choice hypothesis envisages, chromosomes must conjugate in the interphase nucleus. Synapsis cannot, of course, take place along the whole length of the chromosomes (bringing the threat of polyteny), but only on short segments corresponding to Pritchard's regions. In this case the chiasmata, the consequence of crossing-over, must be present in the leptotene stage, although they perhaps cannot be seen under the light microscope. For organisms whose life cycle includes a diploid interphase, such a possibility cannot be ruled out in principle. However, the cytological data are conflicting. In *Lilium,* for example, conjugation of homologues has not been observed before zygotene (Walters, 1968). Moens (1964), on the other hand, considers that in tomatoes pachytene follows immediately after interphase, conjugation takes place not in zygotene, but in interphase, and when the chromosomes first become distinguishable under the light microscope they already have points of contact. Maquire (1967), who studied the mutual arrangement of homologous heterochromatin regions of the chromosomes in *Zea mays* in premeiotic interphase, concluded that they are located much closer together in the nucleus than the heterologous regions. Consequently, conjugation can take place as early as in interphase. However, it could well be that these facts, like the observation that homologous chromosomes are evidently attached by their telomeres in a specific order to the nuclear membrane (Sved, 1966; Engelhardt and Pusa, 1972; Gillies, 1972), simply indicate that in the interphase nucleus there is a distinctive premeiotic associaton of chromosomes which facilitates but does not replace homologous conjugation in meiosis.

Decisive evidence that synapsis takes place in prophase was based on observations made on meiosis in fungi (McClintock, 1945; Singleton, 1953; Lu, 1966, 1967; Rossen and Westergaard, 1966). These plants have no stable diploid interphase state, and they begin meiosis immediately after fusion of the haploid nuclei of the opposed conjugation types. At this moment the chromosomes are in the condensed state characteristic of late prophase in diploid organisms. The homologous chromosomes are spatially separate. It is these condensed and widely separated chromosomes which begin to conjugate. Conjugation, which is essential for recombination, can thus take place in fungi only in meiosis.

The classical investigations (Darlington, 1940) showed that synapsis begins at specific sites of the chromosomes known as contact points. They are usually situated either near the centromere or in the terminal zones. Median conjugation, in which synapsis is initiated in the center of a chromosome arm, is much less frequently observed (see the detailed surveys by Burnham et al., 1972; Maquire, 1972).

As yet nothing is known about the critical moment of initiation of conjugation and the nature of the synaptic forces. The assertion that synapsis is initiated by the random meeting of homologues in leptotene, an event which is favored by the total duration of prophase (Digby, 1910) and association of specific regions of the chromosomes with the nucleoli, chromocenter, and nuclear membrane (Moses, 1968) or the assignment of some form of mystic "pairing urge" to the chromosomes (Darlington, 1937), all shed little light on this problem. Hypotheses which attempt to interpret synapsis from the physical point of view [by forces of electrostatic attraction (Friedrich-Freksa, 1940) or by resonance forces (Jordan, 1941; Fabergé, 1942)] likewise cannot be regarded as satisfactory (for a critical survey, see Vol'kenshtein, 1965). Serra (1947) put forward the idea of elastic fibers connecting homologous chromomeres. Holliday (1968) also postulates the presence of a contractile protein with two active centers, specifically identifying particular homologous nucleotide sequences (for example, contact points). Attachment to them facilitates conformational contraction of the protein and synapsis (see also the zygomere–zygosome hypothesis; King, 1970).

Watson and Crick (1953b) advanced the idea that synapsis may be due to complementary interaction between denatured segments of DNA. However, if such a possibility cannot be ruled out for prokaryotes (Thomas, 1966), in the condensed chromosomes of eukaryotes uncoiling of DNA and Watson–Crick interaction between complementary strands must be difficult. This argument was advanced, in particular, by Pritchard. In his opinion, conjugation must take place before the prophase of meiosis, when the chromosomes are most decondensed, so that local denaturation and the search for and establishment of contacts between the homologues are facilitated. Although there is no doubt at the present time that pairing of complementary DNA strands is essential for crossing-over to take place, the fact that conjugation in fungi is initiated between condensed chromosomes is evidence that there may be two types of pairing: cytologically observable synapsis, essential for bringing the homologous chromomere regions roughly opposite each other (at a distance of the order of 3000 Å, Moens, 1969), and effective pairing at the level of complementary DNA strands.

The spatial organization (topology) of the crossing-over process is based in eukaryotes on a special structure known as the synaptinemal complex, first discovered by Moses (1956) during an investigation of the submicroscopic organization of paired homologous chromosomes (bivalents) in the prophase of meiosis in certain fish.

Further investigations showed that it is found in virtually all species in which crossing-over and chiasma formation normally take place. The synaptinemal complex consists of three parts (survey Moses, 1968) — two axial components (Fig. 53, 3, 4), each of which is arranged along its own chromosome before pairing actually begins, and a central element (Fig. 53, 1, 2), formed during pairing from cross-strands leaving the axial elements. When the cell emerges from prophase the axial elements dissociate from their homologues and are dissolved. These components are protein rods which stretch along the entire chromosome. The cross-strands of the central element also consist mainly of protein, but they do also contain DNA. The relationship between the protein and DNA in the central element is not yet quite clear. It may be that most of the strands are protein in nature (Solari and Moses, 1973), and that they maintain the paired state, but DNA strands must evidently also exist among them — these are evidently only the basic templates: according to some observations fibrils 100 Å in diameter may leave the axial components and return to the same area, forming loops up to 1 μ in length which are broken by deoxyribonuclease.

However, the resulting chiasmata can only be seen, even under the electron microscope, in early diplotene, where they look like short "bites" of the synaptinemal complex (Westergaard and Wettstein, 1970).

The presence of a synaptinemal complex is essential for meiosis to run a proper course (pairing, recombination, and separation). Agents blocking protein synthesis of limiting DNA synthesis in the zygotene stage (4.4) prevent the formation of the synaptinemal complex (Roth and Ito, 1967; De, 1968; Parchman and Stern, 1969). The gene controlling the formation of this complex in *Drosophila* has been identified as a recessive suppressor of crossing-over located in the 3rd chromosome (Gowen, 1933). In females homozygous for mutation in this gene crossing-over is totally suppressed in

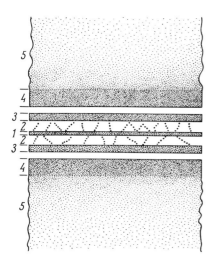

Fig. 53. Structure of the synaptinemal complex (longitudinal section through a bivalent) (from Moses, 1968). 1) Central element: basic and acid proteins (a little DNA?); 2) central space, crossed by protein bands, at least some of which contain DNA; 3) inner elements of the axial complex (DNA, basic proteins, acid proteins); 4) outer elements of axial complex (basic and acid proteins, a little DNA); 5) main mass of the chromosome (microfibrils): DNA, proteins.

all chromosomes, and electron-microscopic investigations show that the formation of a synaptinemal complex in them does not even begin (Meyer, 1964; Smith and King, 1968). Other anomalies of meiosis, such as failure of the chromosomes to separate, are also frequently observed in these females (Hall, 1972). The synaptinemal complex is absent in the *Drosophila* male as well as in other organisms with achiasmatic meiosis, although this correlation is not absolute. Chiasmata are absent, for example, in the scorpion-fly *Panorpa nuptialis*, but a synaptinemal complex has been found (Gassner, 1967; see the survey by Westergaard and Wettstein, 1972). This shows that the presence of a synaptinemal complex is an essential condition for crossing-over (but see Grell et al., 1972), but is not a sufficient condition by itself. The complex is thus responsible neither for pairing itself (for it is absent in paired achiasmatic bivalents and during the formation of polytene chromosomes) nor for reduction (in achiasmatic meiosis in *Drosophila* males it occurs normally), and it is evidently not essential for recombination as such (somatic crossing oven in *c(3)G* females takes place at normal frequency), but it is essential for regular and intensive exchanges. It seems that the complex enables effective synapsis to take place between DNA molecules, it maintains the chromosomes in a paired state for a certain length of time, and it separates the DNA participating in recombination (the basic templates) from the remainder of the chromosomal DNA.

It is evident that this unique equipment must affect the regulation of recombination at both intergenic and intragenic levels. Wettstein (1971) showed that in *Neotiella* each pair of sister chromatids forms only one lateral (axial) component. After the homologues have come together to within a distance of 3000 Å, as a result of some unknown mechanism, the axial element shifts from the space between the sister chromatids outside, i.e., into a position at the side of them both. This explains why conjugation and crossing-over involve only two homologous chromatids at each point. The above-mentioned *c(3)G* mutation in the heterozygous state increases the frequency of appearance of R_1 class both absolutely and relatively in the standard flank system (Carlson, 1972). See also the interesting speculations on this theme by King (1970) and Putrament (1971).

Further progress toward the understanding of synapsis in eukaryotes will evidently be made through detailed electron-microscopic investigations (Comings and Okada, 1970) and the study of mutations specifically influencing this process (Beadle, 1930, 1933; Riley, 1966; Sandler et al., 1968; LaCour and Wells, 1970; Robbins, 1971; Dover and Riley, 1972; Evans and Macefield, 1972; Baker and Carpenter, 1972).

Whether there is anything in prokaryotes analogous to the synaptinemal complex, and whether pairing of complementary strands is preceded by pairing of intact DNA molecules, analogous to conjugation observed cytologically in eukaryotes, or whether it is limited to one stage only (complementary interaction), as Thomas (1966) considers, is not yet known.

4.7. Structural Aspects of Stadler's Modified Polaron Hypothesis

In the earlier sections of this chapter convincing evidence was given that the basic assumptions on which the various alternative forms of the copy-choice hypothesis (the conservative character of DNA synthesis and coincidence between the times of replication and recombination) rested are not confirmed by experimental data. In the next sections further evidence will be given, together with direct proof, that the mechanism of genetic recombination is physical breakage followed by reunion of the chromosomes. Before turning to the examination of these matters, however, let us take a more detailed look at the last of the modifications of the copy-choice hypothesis, that due to Stadler.

It was based on acceptance of experimental facts given in the preceding sections: the replication of DNA in meiosis is semiconvervative in character and it ends before the beginning of meiosis; crossing-over and conversion take place in the prophase of meiosis, when synapsis occurs between homologous chromosomes; the meiotic chromosome is single-stranded, i.e., it consists of a single DNA molecule. All these facts rule out the possibility of a replication mechanism of crossing-over. Crossing-over must therefore take place by breakage of the chromatids at the linkers. So far as conversion is concerned, this is the only possible mechanism of recombination within DNA molecules (polarons), and in order to explain it, the possibility of copy-choice must still be allowed. How can these situations be reconciled? Stadler finds a way out of the dilemma as follows. He postulates that while the main part of the chromosome replicates semiconservatively before synapsis, individual polarons for some reason or other are unable to replicate in the S-period. As a result, during synapsis homologous chromosomes in one region form a three-chromatid structure. As replication of the polaron proceeds to its completion, which it does in a conservative manner, the replica is switched to the homologous chromatid. If the replica does not return to the original template as the result of a second switch (conversion without recombination of the flanks), replication of one of the chromatids also is incomplete. The residual free end of the polaron is unable to initiate a new replication cycle (polarity of replication in the polaron is assumed), and it therefore interacts with the homologous chromatid, giving rise to stress which leads to breakage after completion of replication of the 4th chromatid (conversion accompanied by recombination of the flanks).

The conservative character of DNA replication in meiosis was essential for Stadler to explain the frequency of sister spores mainly observed in his system with conversion (6 : 2 segregation). The presence of postmeiotic segregation (5 : 3) in other systems was ignored. By the time that the modified polaron hypothesis was put forward it had already been shown that during DNA replication *in vitro* the synthesis of two strands of opposite polarity takes place in different directions (Kornberg, 1961); furthermore, there were no grounds for hoping that DNA synthesis in meiosis can differ in its character from that in mitosis. Stadler therefore postulated, after Taylor (1958), that the conservative character of synthesis is simulated by replication of the copy: initially only one DNA strand replicates, but later an additional replication cycle begins, in which the newly synthesized strand is used as template. How the replication cycle (recombination in a given polaron) is completed remained unexplained. A very powerful argument both against Stadler's hypothesis and against its most recent modifications (see 5.3) is that segments which have not replicated in the S-phase possess the specific sequence of nucleotides (see 4.4) and their replication is essential only for conjugation of the chromosomes. Toward the end of zygotene the chromosomes have fully replicated and for that reason pachytene synthesis of DNA is now superfluous (or reparative, as interpreted by Whitehouse; see 5.5).

The discussion on replication mechanisms of recombination can be summarized as follows:

1. If crossing-over takes place by copy-choice between two DNA strands simultaneously (i.e., DNA replication is conservative in character), in order to explain multichromatid exchanges crossing-over between sisters must be assumed, although its existence in meiosis is questionable to say the least.
2. The conservative character of DNA replication is contrary to the experimental data.
3. Replication of DNA and duplication of chromosomes take place in the premeiotic phase (the S-period).
4. Synapsis and crossing-over take place in prophase of meiosis, and not in interphase.

All these facts are evidence that none of the hypotheses so far suggested to explain crossing-over (Belling's, Darlington's) is satisfactory. The attempt to explain recombination on the basis of semiconservative copy-choice requires the introduction of breakages into the scheme. Consequently, the copy-choice mechanism can be used only to explain recombination within the gene. However, here also serious difficulties arise. Different types of segregation in asci (5 : 3) and the difference between the frequencies of appearance of the asci 6+ : 2− and 2+ : 6− in one-point crosses remain unexplained. Moreover, the copy-choice hypothesis which was first assumed to explain the molecular precision of recombination also proved unequal to the task. It remained completely unknown how the replica can switch to the new template at precisely homologous points (nucleotide to nucleotide).

Modern views on the mechanism of enzymic synthesis of DNA (synthesis in the same direction from the 5'- to the 3'-end of short polynucleotide fragments cross-linked by ligase, permitting the simultaneous growth of both chains in a replicative fork: Hosoda and Matheus, 1968; Okazaki et al.,

1968; Kornberg, 1969; Taylor, 1973) completely rule out any possibility of "switches" during replication

The crisis facing the copy-choice hypothesis was resolved by discovery of the fact that recombination in prokaryotes and eukaryotes takes place by a mechanism of breakage and reunion.

4.8. Breakage and Reunion

The preceding discussion has shown that crossing-over in eukaryotes is not connected with replication of the chromosomes and it must therefore take place by breakage of the chromosomes and their reunion. Until recently no direct experimental proof had been obtained of this breakage mechanism of crossing-over in eukaryotes. It is sometimes suggested that Taylor's experiments, which showed that exchanges between sister chromatids can be found on the basis of redistribution of the tritium label in them, could be proof of this kind. It must be emphasized, however, that this fact does not prove that crossing-over is based on a breakage mechanism. In fact, breakages between sister chromatids are essential even from the standpoint of the classical copy-choice hypothesis (1.15). It was Taylor (1965, 1967) who also performed the vital experiment which showed that the distribution of tritium label between labeled and unlabeled chromatids in meiosis in the cricket corresponds exactly to that predicted by the chiasmatype theory. In Taylor's experiments the interval between injection of thymidine-H^3 into the nymphs and fixation was so chosen that chromosomes incorporating thymidine in the interphase before the last premeiotic mitosis were recorded in meiosis. These chromosomes started meiosis with only one labeled chromatid. Reciprocal exchanges between homologous chromatids could be found if labeled and unlabeled chromatids participated in them; consequently, they could be found on the average in half of the exchanges, for the remaining exchanges between two labeled or two unlabeled chromatids are not recorded (Fig. 54). The results showed that the frequency of observed exchanges was exactly equal to that predicted from the number of chiasmata and the absence of exchanges between sisters. For example, the mean frequency of chiasmata for the largest chromosome is 3.67. Since the result of

Fig. 54. Redistribution of tritium label between chromatids in meiosis (after Taylor, 1967): A) no exchanges seen; B) exchanges observed. Labeled chromatids are shown in black.

each chiasma must be the formation of two of four chromatids with an exchange of label, the theoretically expected frequency of the observed exchanges is 1.88 : 2 (since only half of the possible exchanges is recorded). The observed frequency was 0.89.

Taylor's experiment was the culmination of a long series of cytogenic studies providing evidence in support of the chiasmatype theory (see also Peacock, 1968, 1970; Douglas and Kroes, 1969). At the same time, this theory cannot satisfactorily explain why the breakages take place. To do this we must descend to the molecular level.

Even the indirect arguments in support of crossing-over by breakage were strong enough to cause the abandonment of the copy-choice hypothesis, and this was the course followed by Freese (1958). So far as intragenic recombination is concerned, at that time it appeared impossible to explain it without invoking the copy-choice mechanism (4.7). It was also assumed that recombinations in microorganisms have something in common with conversion, because both these processes are characterized by nonreciprocity. However, the analogy is very deceptive, for in the latter case the nonreciprocity characterizes the elementary act of genetic recombination, while in the former it characterizes merely the end result of recombination processes hidden from direct observation. In accordance with this argument recombination in *Drosophila* can also be called nonreciprocal, for only one oocyte is formed in the females during meiosis, so that in each individual meiosis only one recombination product can be detected (apparent nonreciprocity).

The fact that the number of replication cycles of phages T2 and T4 coincides with the number of acts of crossing in the pool is, in the opinion of many authorities (Delbrück and Stent, 1961), evidence in support of copy-choice. The absence of this correlation for phage λ was explained by possible recombination between the vegetative chromosome of the bacteriophage with the bacterial chromosome. It was later observed that the kinetics of formation of recombinants is independent of the medium, although in a richer nutrient medium the yield of phage is ten times higher than in a minimal medium. This indicates that recombination processes are independent of molecular synthesis of DNA (Symonds, 1962). Even in the presence of inhibitors of DNA synthesis (FUDR, KCN, etc.), when the yield of the phages is reduced to 1-3 particles per bacterial cell, the recombination frequency not only is not reduced, but is increased (Symonds, 1962; Anraku and Tomizawa, 1964a; Simon, 1965). The fact that DNA synthesis is unnecessary for recombination has also been demonstrated in transformation systems (Voll and Goodgal, 1961; Bresler et al., 1964d; Bodmer, 1965; Dubnau and Cirigliano, 1973). Not only crossing-over, but also recombinations within the gene take place when DNA synthesis is blocked (Sechaud et al., 1965). Direct physicochemical proof that crossing-over consists of an exchange of material between parental chromosomes has also been obtained in prokaryotes.

§4.8] BREAKAGE AND REUNION 125

It has long been known that fragments of DNA of the parental T-even phages appear among progeny particles. Detailed investigations (Kozinski, 1961; Kozinski and Kozinski, 1964) of fragmentation of DNA of phage T4, using the method of density labeling [with 5-bromouracil (5BU) or N^{15} and C^{13}] and investigation of the DNA of the phage progeny in a CsCl density gradient led to the conclusion that each separate phage particle contains only a small fragment of the parental DNA molecule (5-7% of the entire chromosome). The parental light phage in these experiments was labeled with P^{32}, and the bacteria were grown on heavy medium. On analysis of the DNA of the progeny, phosphorus crossing from the parental DNA was found in a band which coincided with the band of the heavy DNA. If DNA molecules of the progeny were fragmented by hydrodynamic gradients or treatment with ultrasound, the parental label was found in a hybrid band corresponding to molecules of which one strand was formed by light and the other by heavy DNA. Denaturation of the material isolated from this band followed by its centrifugation in a density gradient in the presence of formaldehyde led to disappearance of the hybrid band and to migration of the P^{32} label into the light band of parental DNA. These results show that fragments of parental DNA are present in the DNA of the progeny as single-stranded segments. It has been shown (see page 204) that these segments of parental DNA in the molecule of the progeny are single (continuous) subunits and do not consist of several smaller subunits closely linked together (Fig. 55) (Shahn and Kozinski, 1966). It was a long time, however, before the fragmentation process could be connected directly with genetic recombinations in phage T4 because nothing definite was known at that time about the structure of the phage chromosome (a continuous double helix or a double helix with breakages in the separate strands in a chessboard pattern) or as regards the method of its replication (conservative, semiconservative, or dispersive). For this reason the decisive proof that genetic recombination is the result of breakages and reunions was obtained in experiments on another bacteriophage (λ), for which the semiconservative character of DNA replication had been clearly demonstrated.

Light phage λ c mi and heavy phage λ++ were crossed experimentally (Meselson and Weigle, 1961). The c and mi gene determine the morphology of the sterile plaques. The distance between them is 5 map units, or approximately 30% of the entire λ linkage group. The position of these loci on the

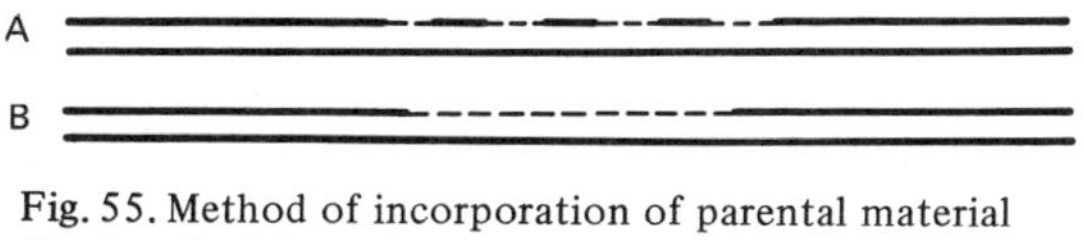

Fig. 55. Method of incorporation of parental material (broken line) into chromosome of the progeny in phage T4 (after Shahn and Kozinski, 1966): A) as discrete portions, B) as continuous subunits.

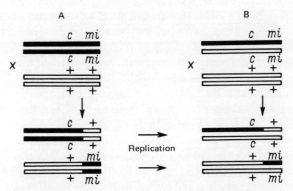

Fig. 56. Distribution of parental material in phage λ recombinants for *c* and *mi* markers: A) recombination before replication, B) after replication. Each strip represents a complementary strand of the DNA molecule (half-chromatid). "Light" strands are shown in black.

map is shown on the diagram in Fig. 56. To understand the map of distribution of the $c\ mi^+$ and $c^+\ mi$ recombinants in the density gradient (Fig. 57) the possible alternatives (A and B in Fig. 56) of their formation on the basis of crossing-over with breakage must be examined in turn.

A. *Original nonreplicated chromosomes recombine.* Since the *c* and *mi* loci are asymmetrically placed on the map, during crossing-over between

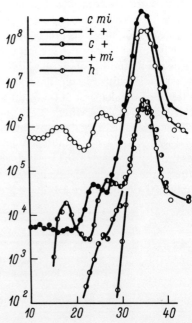

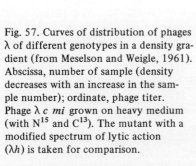

Fig. 57. Curves of distribution of phages λ of different genotypes in a density gradient (from Meselson and Weigle, 1961). Abscissa, number of sample (density decreases with an increase in the sample number); ordinate, phage titer. Phage λ *c mi* grown on heavy medium (with N^{15} and C^{13}). The mutant with a modified spectrum of lytic action (λ*h*) is taken for comparison.

them the $c\ mi^+$ recombinants must inherit 85% of the chromosome of the light parent ($c\ mi$) and only 15% of the chromosome of the heavy (++) parent. Conversely, the $c^+\ mi$ recombinants must inherit 85% of the chromosome of the heavy (++) and 15% of the light parent. Accordingly, the $c\ mi^+$ recombinants appear in the peak roughly corresponding in density to the light phage particles, and the $c^+\ mi$ recombinants in the peak corresponding to the heavy. This was the result which was obtained (Fig. 57). Clearly the peaks of the recombinants are displaced relative to the peaks of the parent phages. The displacement is due to the fact that the recombinants differ in density from the parental phage particles by 10%. This figure is in good agreement with that expected theoretically (15%), but the resolving power of the method is too low to discriminate between crossover and non-crossover transfer of the label (see page 211).

B. Heavy parental phages (++) can replicate before recombination. In this case the c^+mi recombinants will contain only 35% of the label and will occupy a position in the gradient corresponding to hybrid phages. Recombinants formed by interaction between nonreplicating chromosomes will also undergo replication. Clearly this is not reflected in the position of the light recombinants. The heavy recombinants move into the hybrid peak.

Kellenberger et al. (1961) obtained λ mutants ($b2$ and $b5$) which differed considerably in their densities from wild-type phages, probably because they contained deletions amounting to 18% of the entire chromosome. When these mutants were crossed, wild-type recombinants with higher densities than the parental phages and double recombinants with densities lower than that of the parents were obtained. One of the parental phages was labeled with P^{32} and the distribution of the label in the progeny was analyzed. It was found that 88% of the label was bound with the two recombinant classes and only 12% with phages possessing the genotype of the unlabeled parent.

The results obtained in these experiments were completely irreconcilable with the copy-choice model in its pure form, but not in its modified form known as "breakage and copying" (Delbrück and Stent, 1957). According to this hypothesis a fragment lost through breakage of the parental chromosome is restored by replication on the homologous region of the chromosome of the other parent. The choice between breakage – reunion and breakage – copying mechanisms is possible only in experiments in which both parental phages are labeled. The appearance of label from both parents in the recombinants would completely exclude the second mechanism. Two labeling methods were used: biological and physical. A host-induced modification (Ihler and Meselson, 1963) was used as the biological label. Particles of phage λ, grown on *Escherichia coli* P cells (designated λ-P) can infect a culture of another strain *E. coli* C and produce its lysis. However, the progeny liberated during lysis (described as λ-C) is modified by the host cells so that it becomes unable to develop on the original (*E. coli* P) strain. Only

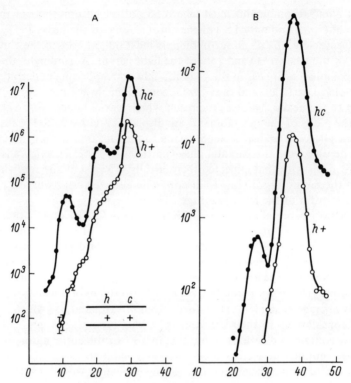

Fig. 58. Curves of distribution of phage λ particles in a density gradient (from Meselson, 1964). A) Distribution of λhc and λh from a $(C^{13}N^{15})$ λ × $(C^{13}N^{15})$ λhc cross. Location of h and c mutations is shown diagrammatically in the bottom right-hand corner. The three maxima in the distribution are formed by phages with nonreplicated, once-replicated, and twice-replicated chromosomes in order of increasing density; B) distribution of λhc and λh from the semiconservative region (fractions 19-21) of gradient A. Fractions 19-21 were isolated from the preceding gradient and recentrifuged. For an explanation of this part of the figure, see page 131. Legend as in Fig. 57.

those particles which obtain at least one complete complementary strand of parental DNA are capable of exhibiting parental specificity. Recombinant particles of these types were found in crosses of appropriately marked phages λ-C and λ-P. Clearly the appearance of such recombinants would be impossible by a breakage and copying mechanism. In another experiment Meselson (1964) labeled one of the parents with N^{15} and the other with C^{13}. One of them was marked with h and c alleles located in the middle third of the genetic map. Recombinants containing a completely labeled chromosome — the result expected only by the breakage and reunion mechanism — also were found in the progeny (Fig. 58).

After it became known that recombination of vegetative phage is controlled by three independent systems, the occurrence of an exchange of material during recombination was demonstrated for each system (Int, Red, Rec) separately (Kellenberger-Gujer and Weisberg, 1971). See page 206 for a detailed discussion.

Similar experiments were carried out on bacteria. In transformation experiments (Bodmer and Ganesan, 1964; Pene and Romig, 1964; Bodmer, 1965) P^{32}-labeled DNA was added to recipient cells growing on a heavy (containing 5-BU) nonradioactive medium. The DNA was then isolated from the cells and investigated in a density gradient. The radioactive label was found to be incorporated into the cell DNA and for this reason it was detected in the heavy peak. Recombinant transforming activity also was found in the same peak. Similar results were obtained in a study of the recombination mechanism during conjugation of bacteria (Siddiqi, 1963; Bresler and Lantsov, 1966; Oppenheim and Riley, 1966, 1967).

All these facts were obtained by the study of recombination between genes. A similar experiment was carried out using heteroallelic strains (Meselson, 1965). The distribution of the parental and recombinant phages by density obtained in this experiment is shown in Fig. 59. The gene within which recombination was studied lies approximately 3.5 map units from one end of the map whose total length is 17 map units. The alleles were intro-

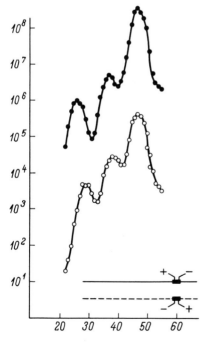

Fig. 59. Distribution of all types of phage particles (black circles) and wild-type recombinants (empty circles) from a $(C^{13}N^{15})$ λ sus 8 $\times$ λ sus 29 cross (from Meselson, 1965). The position of the allelic mutations sus 29 and sus 8 on the map is shown in the bottom right-hand corner of the figure. Legend as in Fig. 57.

duced into the cross so that wild-type phages formed as the result of a single exchange between markers obtained 80% of the chromosome of the labeled (heavy) parent. As Fig. 59 shows, wild-type recombinants were found in three discrete peaks containing: a) 80% of the heavy chromosome (intact, nonreplicated chromosomes recombined), b) 40% (recombination after semiconservative replication of heavy DNA or replication of the recombinant heavy chromosome), and c) 0% (recombination of the light replica of the heavy parent) (see page 208).

After the meaning of fragmentation of the chromosome of the T-even phages, connected with the large number of recombinations which it undergoes in the vegetative pool (if there had been so many crossing cycles in phage λ Meselson would not have been able to find such a simple and clear picture) had become understood, several investigations were carried out on this system and led to an important advance in our understanding of the molecular mechanism of breakage and reunion. As we have just seen, analysis of recombination processes in phages runs into difficulties caused by replication of the recombinant chromosomes. Meselson was able to avoid some of these difficulties by using a high multiplicity of infection (several tens of phage particles per bacterial cell). Under these conditions a few of the phage chromosomes introduced into the cell could avoid replication. In experiments on phage T4 a different approach was used.

It has already been mentioned that neither inhibitors of DNA synthesis nor inhibitors of general cell metabolism (KCN) suppress recombinations in phages, but both may actually increase their frequency two- or threefold. Anraku and Tomizawa (1964b) infected cells with light phage T4 particles labeled with P^{32} (with a multiplicity of unity) and with heavy (5-BU) phages (multiplicity of infection 8), and then incubated the cells in the presence of KCN for 5-60 min. The intracellular DNA was then purified by a method which protected it against breakage, and centrifuged in a density gradient. A DNA peak of intermediate density was found. A complex DNA containing both 5-BU and P^{32} was isolated from this peak and subjected to further purification and analysis. The complex was found not to be an artefact due to aggregation of the molecules, but to possess all the properties of a linear DNA molecule except one: the heavy and light (P^{32}) components of the complex could be separated by heating to the melting (denaturation) temperature of phage DNA. These molecules were described as joint molecules. The complex was treated with ultrasound to break it up into double-helical fragments one-fortieth the size of the original molecules, as a result of which 90% of the P^{32} transferred to the fraction of completely light fragments not containing 5-BU. These results show that the region of union between the two components of the complex is not broken by ultrasound and, consequently, that it must be shorter than one-fortieth of the phage chromosome ($< 3 \times 10^6$ daltons).

Continuing their study of this system (for a summary, see Tomizawa, 1967) these workers used FUDR, which suppresses synthesis specifically (in the absence of KCN), but not completely (leaving 10% of residual synthesis) as the inhibitor of DNA synthesis. Under these conditions the complex dissociates on heating only if it is isolated from the cells during the first few minutes (12-15 min) after infection. After 35-45 min a fraction incapable of dissociating is found. Finally, the joint DNA isolated from the phage particles (still formed in a small quantity in the presence of FUDR) is completely resistant to heating.

These results show that recombination of DNA molecules takes place in (at least) two stages: to begin with two fragments of the parental chromosomes by hydrogen bonding between homologous overlapping single-stranded segments (Figs. 60, 63, 64). Heat-resistant covalent bonds appear later.

How are the "joint" molecules converted into recombinant? Some light is shed on this question by the fact that covalent bonds are not formed in cells poisoned by KCN, in which residual DNA synthesis is on a negligible scale. FUDR does not inhibit DNA synthesis so severely, and it is therefore probable that some slight degree of DNA synthesis is essential for the conversion. This is also confirmed by the fact that although covalent bonds do appear in the presence of FUDR, they do so very late, and it is clear that under normal conditions this process occupies a much shorter time. The need for extra DNA synthesis for the completion of recombination is also clear from the earlier experiments described by Meselson (Fig. 58B). As this figure shows, the peak corresponding in density to the completely labeled recombinants has a small light "shoulder." The half-labeled recombinants and the nonrecombinant heavy phages form distinct peaks. Loss of some of the heavy DNA and its replacement by light DNA synthesized *de novo* are evidently necessary for recombination to be completed. The quantity of

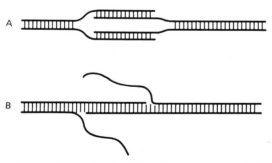

Fig. 60. Diagram showing "joint" DNA molecules (after Anraku and Tomizawa, 1965): A and B) possible alternative forms of union of the fragments.

newly synthesized DNA is 5-10% of the entire chromosome. (The present state of this problem is examined on pages 208-210).

A similar conclusion, that the formation of recombinant molecules takes place in two stages, was also reached by workers investigating transformation (Venema et al., 1965; Ganesan and Buckman, 1968; Dubnau and Davidoff-Abelson, 1971; Dubnau and Cirigliano, 1973) and conjugation (Oppenheim and Riley, 1966, 1967).

The final stage in the work aimed at proving the breakage mechanism of recombination was the development of a technique of electron-microscopic autoradiography (Bresler, Dadivanjan, and Mosevitskii, 1970), by means of which it was possible to observe genetic exchanges in prokaryotes (phage T1) directly. Because of the increased resolving power of this method, accurate information can be obtained on the relative frequencies of crossing-over and semicrossing-over (5.4).

4.9. The Biochemistry of Crossing-over

I have shown in the preceding paragraphs that crossing-over in eukaryotes cannot take place by copy-choice. Moreover, physical breakages and reunions of chromosomes leading to recombination have been found by direct physicochemical methods. How do these breakages and reunions take place? Soon after Meselson and Weigle's discovery it became evident that recombination is a purely enzymic process, and the first proof of this was the fact that synthesis of one of the early proteins induced by phage T4 is essential for the fragmentation of its DNA taking place during recombination. This hypothetical enzyme, whose synthesis was blocked by chloramphenicol in the first few minutes after infection, was called recombinase (Kozinski et al., 1963).

A clearer view of the biochemistry of crossing-over was obtained largely as the result of studies in radiation genetics. The hypothesis linking the lethal action of radiation on the cell with damage to its genetic apparatus (survey: Davies and Evans, 1966) stimulated research into the molecular nature of the injuries included in DNA by radiation. The first investigations showed that the lethal action of UV irradiation is due to the formation of thymine dimers (TT) (Setlow and Setlow, 1962):

A conspicuous result of their appearance is the blocking of DNA replication as was demonstrated by model experiments *in vitro* (Setlow et al., 1963).

Meanwhile, at the beginning of the 1960s, the repair concept was formulated definitively and precisely in radiation genetics (surveys: Val'dshtein and Zhestyanikov, 1966; Korogodin, 1966; Zhestyanikov, 1969). According to this concept, cells possess repair systems which are capable of repairing many of the injuries arising in their genetic apparatus. Differences between parental organisms (or strains) in their radiosensitivity are due primarily to differences in their reparative ability. The phenomenon of photoreactivation was already known and a photoreactivating enzyme capable of splitting dimers in the presence of visible light had actually been isolated. The transforming activity of DNA after exposure to UV irradiation can be partly restored by the action of this enzyme (Wulff and Ruppert, 1962). It was hoped, on the basis of these results, that the differences in sensitivity to UV irradiation could be explained in terms of the ability of different strains to split thymine dimers.

Investigations were carried out on strains of *E. coli*. Setlow and Carrier (Setlow, 1964; Setlow and Carrier, 1964) used strains B_{s-1} and B_r; Boyce and Howard-Flanders (1964) used strains K12 AB 1157 (r) and K12 1886 (s).

It was found that after UV irradiation of the sensitive strain (s) DNA synthesis was severely slowed (the course of DNA synthesis was followed by studying the incorporation of labeled thymidine), and that this was only partially reversible by photoreactivation. DNA synthesis was suppressed by doses inducing the formation of 1-3 dimers per chromosome. UV irradiation of resistant strains (r) led to only moderate suppression of synthesis, which was resumed at the normal rate after a short period. The differences between the sensitivities of the r and s strains are not due to differences in the sensitivity of their genetic structures: the same dose of UV irradiation induces the formation of equal numbers of dimers in both strains. Meanwhile, the energy of visible light is not required for the restoration of DNA synthesis in the r strains and the dimers are not split, as they are during photoreactivation, but they transfer from the DNA into the acid-soluble fraction. The process was therefore called dark reactivation (repair). It was shown that thymine dimers removed from the DNA are parts of oligonucleotides (usually tetranucleotides) with only one terminal phosphate group (pXpUpTpT). After irradiation of resistant, but not of sensitive (paradoxical as it may seem at first glance) cells degradation of the DNA is observed, so that many free nucleotides are found in the acid-soluble fraction along with dimers.

Both groups of workers reached the same conclusion, that dark repair takes place in several stages (Fig. 61). The first stage is endonucleotic exci-

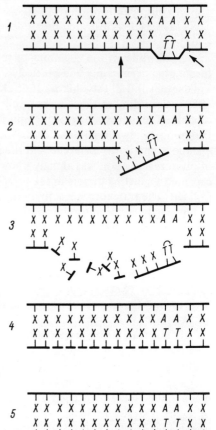

Fig. 61. Diagram illustrating repair of DNA in bacteria after UV irradiation (after Howard-Flanders and Boyce, 1966): T) thymine, A) adenine, X) any base (including T and A). 1) Part of a double-helical DNA molecule containing a thymine dimer (region of primary excision is marked by arrows); 2) excision of dimer; 3) subsequent exonucleotic degradation; 4) reparative DNA synthesis; 5) reconstructed double-helical DNA molecule.

sion of thymine dimers from DNA. The second stage is exonucleotic degradation of the DNA (widening of the gap by successive removal of terminal nucleotides), and this is followed by reparative DNA synthesis. The last stage is that of formation of the final phosphoester bonds, i.e., repair of the single-stranded breakages. The process of dark repair is undoubtedly enzymic in character. Detachment of dimers can take place in the absence of DNA synthesis, under conditions of amino acid deprivation, and in the presence of a small quantity of chloramphenicol sufficient to suppress normal protein synthesis. These facts also show, in particular, that repairing enzymes are present in the cells before irradiation.

Further investigation showed that the excision of injuries (the first stage of repair) consists of three consecutive reactions: identification of the injury in the DNA, production of the endonucleotic "nick" in one polynucleotide chain in the immediate vicinity of the injured segment and of the exonucleotic "cut." Genetic blocking of either of these stages renders the cell incapable of excising the dimers, and the bacteria become sensitive to ultraviolet irradiation (uvr^-) (Bridges and Munson, 1966; Badley, 1968; Moseley, 1969). It was also discovered that not only thymine dimers, but also other abnormal products in DNA induced either by UV irradiation (T-C and C-C dimers) and by other agents (mitomycin C, bifunctional alkylating compounds, and nitrous acid), can effectively be excised from DNA. The identification reaction is controlled by UV-specific endonuclease II (Friedberg and Goldthwait, 1968; Setlow et al., 1969; Mahler et al., 1971; Brent, 1972). It is also responsible for the primary break on the 5'-side of the dimer. This stage is controlled in phage T4 by the *v* gene (Yasuda and Sekiguchi, 1970a,b; Friedberg, 1972) and in *E. coli* by at least four genes: *uvrA, B, C,* and *D* (Taketo et al., 1972). The $uvrD^-$ form differs from the $uvr(A,B,C)^-$ forms in degrading its own DNA. The *rep* mutation (Ogawa, 1970) is located near to *uvrD*. A UV-specific exonuclease participates in the exonucleotic excision of the dimer (Kaplan et al., 1969; Kushner et al., 1971; Ohshima and Sekiguchi, 1972). If these two enzymes act together, besides the dimer they also cut out four or five neighboring necleotides, so that a primary gap in one strand is formed (Grossman et al., 1968).

The destruction of DNA is accompanied by simultaneous reparative synthesis (Haynes, 1966), with the consequent restoration of the double-helical DNA structure.

Evidence for reparative synthesis was obtained in the following experiment (Pettijohn and Hanawalt, 1964). Cells of *E. coli* 15 $T^-A^-U^-$ (deficient in thymine, arginine, and uracil) were irradiated with UV light and then incubated in a medium containing 5-BU-H^3. Incorporation of the isotope into DNA could be judged from the change in the buoyant density of the DNA in a CsCl gradient. In semiconservative DNA replication BU incorporation into unirradiated cells leads to the formation of DNA with hybrid density. During incubation of the cells after irradiation incorporation of BU-H^3 into DNA also was observed, but the DNA isolated from the irradiated cells occupied a position in the gradient corresponding to DNA of normal density. The DNA isolated from the H^3-peak was treated with ultrasound, denatured, and then again investigated in the density gradient. It was found that the BU-H^3 entered both strands of the DNA and was distributed at random in short segments of the chromosome, as would be expected if repair of individual segments of the chromosome took place in the cells during incubation. Reparative synthesis of DNA has also been found in the cells of eukaryotes (survey: Dalrymple et al., 1968; Clarkson and Evans, 1972; Trosko and Wilder, 1973). It is interesting to note that fibroblasts of the skin from patients with xeroderma pigmentosum repair UV-induced injuries much less completely than normal fibroblasts (Cleaver, 1968). They cannot excise pyrimidine dimers (Setlow et al., 1969b). Reparative synthesis of DNA in bacteria is catalyzed by Kornberg's classical DNA-polymerase, coded by the *polA* gene

(Kelly and Whitfield, 1971), although it is not responsible, as was hitherto considered, for molecular replication of DNA (Gross and Gross, 1969; Lucia and Cairns, 1969).

In this case it is carried out by DNA-polymerase III, controlled by the *dnaE* gene (Gefter et al., 1971; Husslein et al., 1971). A DNA-polymerase II, controlled by the *polB* gene (Moses and Richardson, 1971; Campbell et al., 1972; Hirota et al., 1972) has also been found, but its function is unknown for the *polB*⁻ mutants are indistinguishable from the wild type both in their sensitivity to irradiation and in their powers of recombination. This is strange, for yet another polymerase ought to exist and to take part in repair, because *polA*⁻ cells are capable of repairing more than a hundred dimers per gene (Monk et al., 1971). According to Monk et al. (1971) a possible candidate for this role is the product of the *recA* gene, in view of the fact that the double mutant *recA*⁻*polA*⁻ is nonviable. However, *recB(C)*⁻*pola*⁻ is also nonviable. Another candidate is the product *uvrE* (Mattem, 1972). Polymerase I inserts only a few nucleotides for each excised dimer. The second DNA-polymerase, which functions in *polA*⁻ cells, inserts up to 3000 nucleotides (Cooper and Hanawalt, 1972a,b). This polymerase is ATP-dependent (Masker and Hanawalt, 1973). Reparative synthesis in phage T4 is also carried out by polymerase I (George and Rosenberg, 1972).

Boyce and Howard-Flanders (1964) observed a striking analogy between dark repair and genetic recombination. In the latter, as in the former (4.8), breakages in individual chains (strands) of the DNA, partial destruction of the DNA in the region of the gap, reparative synthesis of the DNA, and reunion of the phosphate bonds are evidently necessary. Particular enzyme systems may thus participate simultaneously in the two processes and, consequently, correlation ought to be found between sensitivity to UV irradiation and the disturbed power of recombination (*rec*⁻). Such a correlation was found for the first time in phage T4 (Harm, 1964). A mutation in the *x* gene determining UV sensitivity of the phage reduced the recombination frequency in the *rII* region by 3.5 times. However, the search for the *rec*⁻ phenotype among strains of *E. coli uvr*⁻ gave negative results. This correlation was found to exist only in strains sensitive to both UV and x-ray irradiation. Ray et al. (1971) and Mortelmans and Friedberg (1972) showed that a mutation in the *x* gene, as also in the analogous *y* gene, leads to an increase in sensitivity not only to UV rays, but also to EMS and to x-rays (see also Maynard-Smith and Symonds, 1973; Symonds et al., 1973). In order to isolate *rec* mutants the procedure adopted nowadays is therefore to use sensitivity to any agent inducing single-stranded breakages in DNA as the selective characters (x-rays, methyl methanesulfonate, etc.).* The *rec*⁻ mutants were initially isolated

---

*The nature of injuries in DNA induced by these agents was established at the same time (Szybalski and Lorkiewicz, 1962; Strauss et al., 1968; survey: Zhestyanikov, 1969). Repair of single-stranded

from *E. coli* (Clark and Margulies, 1965; Howard-Flanders and Theriot, 1966; Hertman, 1967; Emmerson, 1968), and later from other bacteria (Searashi and Strauss, 1965; Okubo and Romig, 1966; Prozorov, 1966; Holloway, 1966; Prozorov et al., 1968; Botstein and Matz, 1970; Caster and Goodgal, 1972; Setlow et al., 1972) and eukaryotes (Holliday, 1965, 1966a; Zakharov and Kozhina, 1968; Jansen, 1970; Fortuin, 1971; Hunnable and Cox, 1971; Roth and Fogel, 1971; Rosen and Ebersold, 1972; Smith, 1972). A sharp decrease in the frequency of chiasmata, correlated with injury to the repair system, has also been found during spermatogenesis in sterile men (Pearson et al., 1970, cited by Swietlinska and Evans, 1970). Conjugation of *rec*⁻ mutants of *E. coli* is possible and they can receive the donor's genetic material, but recombinants are formed at a frequency $10^2 - 10^5$ times lower (depending on the strain) than in *rec*⁺ forms.

The nature of the injuries in *rec*⁻ bacteria began to be understood as a result of the study of the metabolism of their DNA before and after UV irradiation (Howard-Flanders and Boyce, 1966). The content of radioactivity in the acid-soluble fraction was determined in cells labeled with tritiated thymidine after various periods of incubation. The radioactivity in wild-type cells was found not to transfer to the acid-soluble fraction during control incubation, but 2 h after UV irradiation up to 7% tritiated thymidine was found in it (470 nucleotides for each thymine dimer removed). In cells of the radiosensitive strain AB 2463 (*rec-13*) irradiation is followed by very intensive DNA destruction (27,000 nucleotides per dimer). During normal growth the cells of this strain spontaneously destroy up to 30% of their own DNA, so that, to use a pun, they have been called "reckless." Another strain of *E. coli*, possessing the mutation *rec-21* which is mapped in a different region of the chromosome from *rec-13* (*recB* between *thyA* and *argA*), possesses sensitivity to x-rays which is intermediate between the wild type and *rec-13*. No spontaneous DNA degradation has been found in it and destruction of its DNA after irradiation is actually less than in the wild type (270 nucleotides per dimer), and it has accordingly been named "careful." An even more "prudent" mutant, *rec-22* (region *recC*, closely linked with *recB*: Emmerson, 1968), destroys only 25% of the amount of DNA degraded by the wild-type strain after irradiation.

---

breakages takes place with the participation of the same enzymes as in the case of UV injuries, with the exception of those which operate at the excision stage (Howard-Flanders and Boyce, 1966; Cleaver, 1971). For this reason it is virtually impossible to distinguish between *uvr*⁻ cells and the wild type as regards their sensitivity to x-rays. Most breaks are healed by polymerase III (Youngs and Smith, 1973) and polymerase I (Town et al., 1971; Laipis and Ganesan, 1972; Pauling et al., 1972; Jacobs et al., 1972) so quickly that they cannot be detected by the use of the technique of McGrath and Williams (1966). This process takes place even in a buffer and it is accompanied by the insertion of 1 to 3 nucleotides (Painter and Young, 1972; Worthy and Elper, 1972). Any residual single-stranded breaks are repaired by the slow *recA*/*exrA* system (Kapp and Smith, 1970; Sedgwick and Bridges, 1972), either by inducing exchanges (see below) or without them (Bridges, 1971).

Why do the *rec*⁻ strains behave in this way? The nature of the injury in *rec-13* is evidently determined, not by loss of any of the repairing enzymes by mutation, but rather by regulation of their interaction. The nuclease and reparative polymerase must act in strict coordination in the wild strain so as to ensure minimal, but not permit excessive DNA degradation. This regulation can take place through a special exonuclease regulator, normally restraining its activity or watching to ensure that the exonuclease is replaced by polymerase exactly at the right time. The regulator in *rec⁻A* (the reckless mutant) has evidently lost its functions. Large areas of the chromosome in this strain must have become "exposed" (transferred into a single-stranded state) in the process of repair, as has been demonstrated experimentally (Smith and Ganesan, 1968).

Exonuclease V is responsible for this destruction (see below and Youngs and Bernstein, 1973). A protein coded by *recA* is an inhibitor of this enzyme (Csordas et al., 1972). The *lex* and *zab* genes are evidently related to regulation of the *recA* function (Castellazzi et al., 1972; Mount et al., 1972; Moody et al., 1973), but they have no effect on recombination.

Converging processes of degradation on the complementary strands lead to fragmentation of the chromosome and death of the cells (Pollard et al., 1966). In the *rec-13* strain (injury in the *recA* gene, located between *cysC* and *pheA* markers) the surviving recombinants (0.001% of the control) are not true recombinants but merodiploids (Low, 1968). Integration of the *F* factor in the *recA*⁻ chromosome also takes place 2500 times less frequently (DeVries and Maas, 1971). Inability to form recombinant chromosomes is evidently due to the fact that during recombination any two consecutive gaps on complementary strands (Chapter 5) are separated by a distance shorter than the "critical" and this causes death of the cell.

In *rec⁻B* and *rec⁻C* (careful mutants) ATP-dependent nuclease is inactivated (Buttin and Wright, 1968; Willetts and Clark, 1969; Oishi, 1969; Goldmark and Linn, 1970; Barbour and Clark, 1970), called endonuclease (Wright et al., 1971), but also possesses ATP-stimulated endonucleotic activity on single-stranded DNA (Goldmark and Linn, 1972; Nobrega et al., 1972). This enzyme belongs to the leapfrog type for it removes oligonucleotides 7-9 bases long. It does not act on nicks in DNA but only on double-helical ends, leaving behind it exposed single-stranded regions. The gene *recB* is a structural gene of this enzyme; *recC* either regulates its synthesis or modifies the protein of the *recB* gene, converting it into the enzyme (Tomizawa and Ogawa, 1972). Similar enzymes have also been isolated from other bacteria (Vovis and Buttin, 1970; Dorp, 1972; Friedman and Smith, 1972; Prozorov et al., 1972).

The ATP-dependent nuclease from *H. influenzae* begins by inflicting "staggered" single-stranded bites over a distance of 2000 bases and carries out denaturation of that region, with the result that double-helical fragments surrounded by single-stranded tails 2000 bases long are formed. These fragments are then degraded down to oligonucleotides (Friedman and Smith, 1973). The surviving recombinants (0.1% of the control) are completely normal and, in addition, the recombination frequency of unselected markers among the surviving recombinants is indistinguishable from the control. In a similar way the frequency of crossing-over in many radiosensitive strains of fungi with blocked meiosis, determined in the surviving progeny, likewise remains unchanged (Chang and Tuveson, 1967; Haefner and Howrey, 1967; Snow, 1967).

Further investigations showed that Low's conclusion is valid only for the inner regions of the merogenote. At the proximal (and also possibly, at the distal) end *rf* falls sharply (from 0.52 to 0.092) in all nine independently isolated *recB(C)⁻* strains (Haan et al., 1972). The reduction in recombination at the ends, however, is insufficient to explain the decrease in the total yield of recombinants to 0.1%. A sufficiently high level of recombination is evidently formed (Barbour, 1972) in strains without ATP-dependent nuclease by means of an alternative pathway inducing exchanges in the middle of the merogenote (Haan et al., 1972).

This pathway has in fact been found. Revertants from *recB⁻* (or *recC⁻*) strains have their recombination properties restored by mutations in two suppressor loci *sbcA* and *sbcB*. In double *sbcA⁻recB(C)⁻* mutants with the *rec⁺* phenotype an ATP-independent nuclease appears instead of an ATP-dependent nuclease (the *recE*-pathway; Barbour and Clark, 1970). The *sbcA⁻*-regulatory mutation, derepressing the *recE* structural gene, is called *rac* by Low (Templin et al., 1972). It is postulated that this is not a bacterial gene but the *redX* gene of a defective prophage (Low, 1973). As well as exonuclease V, the *sbcB⁻recB(C)⁻* strains were also deficient in exonuclease I (Barbour et al., 1970), acting on single-stranded DNA from the 3'-end. The pathway activated by the absence of exonuclease I was called the *recF* pathway (Kushner et al., 1971). Three genes have been identified in this pathway: *recF*, *recL*, and *recK*, mutations in which block recombination in *sbcB⁻recBC⁻* strains only. The *sbcB⁺* strains thus recombine by means of the *recBC* pathway (since exonuclease I blocks the *recF* pathway, in which the 3'-ends are evidently the intermediate structures). The *sbcB⁻* strains can recombine equally successfully by either pathway, and the double mutants *sbcB⁻rec(B,C,F,L* or *K)⁻* are phenotypically *rec⁺*, for a decrease in recombining activity by 50% passes unnoticed. These strains are capable of transformation (Oishi and Coslov, 1972a,b; Wackernagel, 1973). It is interesting to note (Unger et al., 1972) that another mutation inactivating exonuclease I (*xonA⁻*) does not lead to suppression of the *recB(C)⁻* phenotype. Meanwhile the resistance of *recB(C)⁻* to ultraviolet radiation is completely restored in the presence of either of these mutations (Kushner et al., 1972). This points to the presence of another activity of exonuclease I which differs from its 3'

→ 5' nuclease action. The *recFKL* genes evidently also determine a certain nuclease, for the antinuclease of *recA* is included in both pathways (see the surveys by Clark, 1971, 1972).

It is an extremely interesting fact that mobilization of the chromosome (Wilkins, 1969), recombination between the markers on *F* and the chromosome (Hall and Howard-Flanders, 1972), rescue of Lac^+ from the UV-irradiated episome (Cole, 1971), recombination between *Flac* episomes (Willetts, 1972) and, evidently, recombination within the gene (Zlotnikov and Khmelnitskii, 1973) in $recB(C)^-$ strains (unlike $recA^-$) take place almost as efficiently as in the wild type. The mechanism of semicrossing-over in these strains is evidently not damaged. Several more *rec* mutations have been discovered in *E. coli* in genes with as yet unknown functions (Lloyd, 1972); *def* (Martin et al., 1969), *cetC* (Holland and Threlfall, 1969), and *recD,G,H* (Storm et al., 1971), together with an *Hfr*-specific protein stimulating recombination (Goldfarb et al., 1973).

Activity of exonuclease V disappears in *E. coli* cells infected with T4 or λ phages (Hobom and Hobom, 1972). In the second case inhibition of the enzyme is linked with the functioning of the γ gene. It is clear that the *RecE* pathway is not inhibited by this gene (Unger et al., 1972; Unger and Clark, 1972) (see page 209). The generalized recombination system of phage λ (the *Red* pathway) incorporates (besides the γ gene) at least two other genes: *red*α (evidently the analogue of *recB*) and *red*β, the precise function of which is unknown, for *in vitro* the λ-exonuclease coded by *red* (Signer et al., 1968; Radding, 1970) acts equally effectively both as a complex with the β-protein and without it (Cassuto and Radding, 1971). The reaction carried out by this enzyme *in vitro* (Radding and Carter, 1971; Carter and Radding, 1971) consists essentially of the assimilation of one strand. Like exonuclease V, the enzyme does not act on nicks in double helices, but in the case of superfluous linkage (Fig. 60B) it begins to destroy one strand of the double helix, thus enabling the excluded (superfluous) part of the complementary strand to continue to pair until it has all been assimilated and the excess has disappeared. A single-stranded break is the stop signal for this enzyme. Using λ exonuclease, Cassuto and Radding (1971) carried out an experiment closely resembling recombination *in vitro* (Fig. 62). Tritiated DNA of phage λ was divided into *l* and *r* strands. DNA of the same phage, labeled with P^{32}, was broken in half and the resulting fragments also were divided into separate strands. The whole *l* strands were joined first with the *r* fragments and then with the whole *r* strands. The structure of the ring chromosome of phage λ was formed with a superfluous single-stranded fragment equal in length to half of the DNA molecule. This structure was supported by hydrogen bonds. After addition of the *red* enzyme, the P^{32}-labeled nucleotides were cut out and a double-stranded ring with the characteristic sedimentation constant was formed.

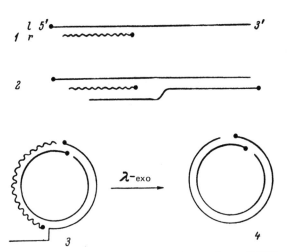

Fig. 62. Scheme of Cassuto and Radding's experiment (1971). The dot marks the 5'-end of the DNA strand, the straight line the H^3-labeled DNA strand, and the wavy line the single-stranded P^{32}-labeled DNA fragment: 1-4) sequence of events. Arrow points to moment of addition of endonuclease λ.

In order to understand the nature of recombination systems it is important to compare the functions of bacterial *rec* genes and of phage *red* genes. These systems are evidently interchangeable, for the *rac* locus restores the ability of *E. coli rec⁻* merozygotes to form recombinants (Low, 1973).

The existence of different types of *rec⁻* mutants shows, first, that during recombination endonucleotic gaps actually arise in the polynucleotide strands and induce a process of degradation; second, that for recombination to take place a certain minimum of exonucleotic DNA degradation, initiated by these breakages, is essential; third, that the process of DNA degradation must take place within certain reasonable limits, for otherwise the converging processes of degradation on the complementary strands would lead to fragmentation of the chromosome. Further information on the enzymes participating in recombination has been obtained by the study of conventionally lethal mutants of phage T4.*

---

*Conventionally lethal mutations of two types are known. The first type includes "opal," "amber," and "ochre" mutations. They are connected with the appearance of nonsense codons in the genetic text, leading to termination of the translation process (synthesis of the polypeptide chain). In some strains of *E. coli* there are suppressor genes which code a modified transfer RNA enabling translation to be resumed. These mutants can reproduce normally on these (permissive) Su (*am*) and Su (*ochre*) strains. The second type of conventionally lethal mutations includes thermosensitive mutants (*ts*). These *ts* mutations have a protein structure which is so modified that they are unable to function at raised temperatures (42°C). At ordinary temperatures the *ts* mutants of the phages develop normally. The *rII* mutations which can grow only on *rex⁻* mutants of *E. coli* B can also be included among the conventionally lethal group (Gussin and Peterson, 1971).

No true *rec⁻* forms have been isolated from phage T4, only partially defective forms Ende and Symonds, 1972). The reason may be that T4 (and also P22, Russell and Botstein, 1972) replicates only after recombination-dependent circularization (Mosig et al., 1972). The main results with this system have therefore been obtained by studying control over the formation of joint molecules and their conversion into recombinant. The first requirement for their formation is the enzyme "nickase" (endonuclease II, Kozinski and Felgenhauer, 1967; Tomizawa, 1967). The action of the endonucleases of genes *46* and *47*, producing single-stranded gaps at the sites of the primary breaks, is also necessary. In this respect the nucleases of T4 differ from the $\lambda red\alpha\beta$ complex and the exonuclease V of *E. coli* (Prashad and Hosoda, 1972). If these enzymes are defective there is a sharp decrease in the frequency of recombinatons (Friedberg and Goldthwait, 1968; Sadowski et al., 1968) and the "branched" molecules disappear (Broker and Lehman, 1971). The fact that the joint molecules of T4 actually possess the structure illustrated in Fig. 63 (3), i.e., that they contain single-stranded gaps of the order of several hundreds of nucleotides on both sides of the overlap, was proved by the experiments of Anraku et al. (1969), who showed that they can be converted *in vitro* into recombinant forms only by the combined action of DNA polymerase of T4 and ligase (see below), but not by either of these enzymes separately. Parental *T4pol⁻lig⁻* DNA extracted from cells contains many gaps and single-stranded breaks, but the recombination frequency is increased, joint molecules are formed, and branches increase in number (Broker and Lehman, 1971).

The gene (*32*) controlling the formation of joint molecules (Tomizawa et al., 1966) has been identified. Despite the fact that synthesis of none of the known enzymes participating in replication is disturbed in this mutant, it can neither synthesize DNA nor form joint molecules. One possible hypothesis regarding the function of gene *32* was that it controls a protein which facilitates separation of the polynucleotide chains necessary both for recombination and for replication. In fact, a protein inducing a hyperchromic effect and binding cooperatively with single-stranded DNA (Carrol et al., 1972) has been isolated from the thymus (Phillips, 1968), from *Pneumococcus* (Kohoutova et al., 1970), from *E. coli* cells infected with phage T4 (Alberts and Frey, 1970), filamentous bacteriophages (Oey and Knippers, 1972; Alberts et al., 1972), and uninfected *E. coli* cells (Sigal et al., 1972). Protein of gene *32* of phage T4 denatures DNA at a physiological temperature in regions rich in A-T, but not at sites of single-stranded breaks (Delius et al., 1972).

The last stage in the long chain of enzyme reactions taking place during repair and recombination is the uniting of the polynucleotides with each other by phosphoester bonds. An enzyme (polynucleotide-ligase: Weiss and Richardson, 1967; silase: Gellert, 1967; Zimmerman et al., 1967; poly-

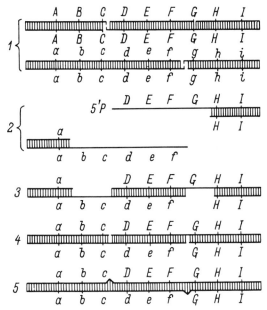

Fig. 63. Enzymic stages of recombination (from Szybalski, 1964): 1) double-helical gaps in recombining chromosomes do not occur at precisely homologous sites (in the upper helix between C and D markers, and in the lower between f and g); 2) exonucleotic degradation of fragments exposing terminal nonoverlapping single-stranded segments; 3) formation of a joint molecule; 4-5) repair of the gaps and restoration of the phosphoester bonds.

nucleotide-joining enzyme: Olivera and Lehman, 1967), catalyzing this reaction, has now been isolated and studied. Ligase takes part not only in reparative (Badley, 1968), but also in ordinary (molecular) synthesis of DNA (4.7). The gene (*30*) controlling the synthesis of ligase has been identified in phage T4 (Fareed and Richardson, 1967). Phages with the *ts* mutation in this gene do not yield a viable progeny at 42°C. Analysis in an alkaline sucrose gradient shows that the newly synthesized DNA splits up into single-stranded fragments. It has also been shown that the presence of phage-specific ligase is an essential condition for the conversion of joint into recombinant molecules.

The double mutant *rII*(*30*) can grow on *E. coli lig*$^+$ (Carlson and Kozinski, 1970), but not on *E. coli lig*$^-$ (Gellert and Bullock, 1970). Krisch et al. (1972) showed that DNA of phage *rII*(*30*) has multiple breaks, repaired slowly by the host's ligase, and *rf* is increased by four times. This effect is absent, however, in *lop* strains of *E. coli,* producing an excess of ligase. The product of the *rII* gene is evidently a membrane protein (Weintraub and

Franklin, 1972; Krylov, 1972). Mutations in genes *44* and *59* (Davis, 1972) likewise affect recombination in phage T4 (Davis, 1972).

Recombination and repair enzymes appear in a strictly coordinated fashion in meiosis (Hotta and Stern, 1971b; Howell and Stern, 1971) and are induced by irradiation in mitosis (Holliday, 1971).

The following principal enzymic stages in recombination can thus be distinguished (Fig. 63):

1. Double-stranded (double-helical) nicks (endonuclease I) or single-stranded nicks in DNA generated by endonuclease II (nickase) (Hurwitz, 1967);
2. Dissociation of complementary strands of DNA (or exposure of single-stranded segments) of double-helical fragments (Kozinski and Felgenhauer, 1967) by terminal excision of mononucleotides from the 3'-OH end (endonuclease III);
3. The formation of joint molecules; it has been shown (Alberts, 1970) that sorption of DNA-binding protein simultaneously facilitates renaturation also; stages 3 and 4 can be catalyzed simultaneously by λ exonuclease in the strand assimilation reaction;
4. Reconstruction of the double-helical DNA molecule by exonucleotic degradation and reparative synthesis (enzyme of the DNA-polymerase type, but not Kornberg's DNA-polymerase; Gross and Gross, 1969), which successively fills in the gap with nucleotides from one end and uses the intact complementary strand as the template; hybrid DNA formation may also result from rotatory diffusion (Meselson, 1972);
5. If the 3'-end of the polynucleotide chain is phosphorylated, 3'-deoxynucleotidase (phosphatase) dephosphorylates it (3'P–3'OH); the 5'-end is dephosphorylated by transfer of phosphate from ATP to the 5'-OH end of the polynucleotide chain by the enzyme 5'-hydroxylpolynucleotide-kinase; the recombinant molecule is finally formed by establishment of phosphoester bonds (an enzyme of the ligase type reunites the cuts in each chain).

4.10. The Mechanism of Induced Recombination

Direct methods of studying the mechanism of crossing-over did not appear until the 1960s, with the development of isotope techniques (4.8). Previously investigators had to be content with indirect approaches. In particular, it was hoped that the study of the effect of external factors on crossing-over would shed light on the mechanism of this process. Various factors, ranging from temperature to carcinogens, were used for this purpose. For work on *Drosophila* the agent mainly used was ionizing radiation, the effect of which on crossing-over had been discovered in 1923 (Maworth and Swenson, 1923). The results of these investigations were rather contradictory (4.5) and, apart from the basic conclusion that ionizing radiation increases the frequency of crossing-over (at least in regions near the centromere in the case of irradiation at the pachytene stage), no progress toward the understanding of this effect was achieved (surveys: Whittinghill, 1955; Zakharov and Inge-Vechtomov, 1961). During this period of mechanistic interpretation of spontaneous crossing-over, the most natural suggestion was that induction is connected with an increase in the frequency of breakages in the chromosomes (Muller, 1932). In fact, ionizing radiation induces not only crossing-over, but also various chromosome and chromatid aberrations, which are caused, in the classical view (Stadler, 1928; Lea, 1955), also by breakages and subsequent reunions of the chromosomes. However, facts contradicting the breakage hypothesis had been obtained (Yost and Benneyan, 1957). The most substantial objection was that the frequency of crossing-over is increased by the action of x-rays only in regions near the centromere, whereas in distal parts of the chromosomes crossing-over is suppressed. Other factors such as temperature, age (Grell, 1966), and inversions (the Schultz–Redfield effect), which do not induce chromosomal aberrations, also have a similar action on crossing-over. Another essential fact was that during induction of chromosomal aberrations both viable (reciprocal: X-exchanges) and nonviable (U) exchanges are formed, whereas during crossing-over only the first type of exchanges is observed. Accordingly the most popular hypothesis was that of Matsuura (1940), who postulated that the increase in recombination frequency is linked with uncoiling of the heterochromatin regions near the centromere (i.e., an increase in the distance between the gene), accompanied by compensatory coiling of distal regions of the chromosome.

The hypothesis of direct breakages (fractures) of chromosomes was also rejected an an explanation of the induction of chromosomal aberrations. A new conception of potential injuries (surveys: Evans, 1962; Luchnik et al., 1964) appeared, according to which irradiation induces not true breakages in chromosomes, but injuries of an unknown nature which can either be repaired or manifested as chromosomal aberrations. As the mechanism of manifestation it was suggested that a nonhomologous crossing-over takes place, either between different chromosomes or between different parts of the same chromosome (chromatid). This possibility was first mentioned by Belling (1927) and Serebrovskii (1929). The hypothesis of nonhomologous crossing-over was for a long time regarded as unrealistic, for it assumed that a single "hit" was sufficient to induce reciprocal aberrations. At the same time, their frequency was shown to be dependent on the square of the dose.

In some brilliant work to study the relative numbers of different types of chromatid aberrations in *Vicia faba,* Revell (1956, 1959) showed that the hypothesis of nonhomologous crossing-over completely explains all the facts if two important reservations are made. First, as Anderson (1936) postulated, nonhomologous crossing-over must take place at the four-chromatid stage, i.e., after replication of the chromosomes; second, the process can only be induced if two independently arising injuries are present. Within the framework of Revell's contact-recombination hypothesis the analogy between induced crossing-over and induced chromosomal aberrations was emphasized still more,* but the question of the mechanism of crossing-over still remained completely open. A solution to the problem had to await a concrete explanation of the nature of the potential injuries and elucidation of the mechanism of spontaneous crossing-over.

The study of the effect of various agents on recombination was not limited purely to meiotic systems. Soon after the discovery of mitotic recombination in fungi it was shown that it can be very effectively induced by a wide variety of agents (1.16). Research in this field was conducted mainly with ultraviolet irradiation, for the equipment is easier to handle than the x-ray apparatus.† A number of general principles were established (surveys: Zimmermann and Schwaier, 1967; Holliday, 1968):

(1) The frequency of mitotic recombinations is increased 100-1000 times. (2) With the same dose of irradiation, conversion is induced 10 times more effectively than crossing-over. (3) Both types of genetic recombinations are induced most effectively at the stage of the cell cycle which is maximally sensitive to irradiation (this stage precedes or coincides with the stage of DNA synthesis in the cell). (4) Both chromosome and gene maps, in good agreement with meiotic maps, can be drawn on the basis of recombination frequencies. (5) Mitotic recombinations are effectively induced by caffeine, fluorodeoxyuridine, mitomycin C, and various mutagens and carcinogens (these agents also induce recombination in meiosis). Since a common effect of virtually all these agents (besides the induction of recombination) is a temporary depression of DNA synthesis, it was postulated that depression of DNA synthesis leads to a state of unbalanced growth, which induces pairing of chromosomes and recombination (Holliday, 1964a). No more concrete hypotheses were formulated until recently (survey: Putrament, 1967a).

---

*The common nature of the mechanisms of formation of chromosomal aberrations and crossing-over is emphasized by the observations of Schacht (1958). In his experiments on *Drosophila* males (in which crossing-over does not normally take place) irradiation of the cells at presynaptic stages induced crossing-over, but at postsynaptic stages it induced chromosomal aberrations (translocations). The breakages arising in the chromosomes evidently lead to homologous exchanges (crossing-over) if synapsis is present, but to nonhomologous exchanges (translocations) in its absence. It is also remarkable that spontaneous breakages take place in the region of chiasma formation in species with localized chiasmata (Walters, 1956).

†Unfortunately it is difficult to use UV-irradiation in experiments on *Drosophila* and similar objects because of its low penetrating power (Browning and Altenburg, 1964; Proust and Prudhomme, 1968).

The history of the study of induced recombination in prokaryotes is a good illustration of the end result of attempts to explain one unknown by means of another: one of the arguments in support of the copy-choice mechanism for a long time was the fact that the recombination frequency is increased by UV irradiation of phages or bacteria. Jacob and Wolman (1962) put forward the hypothesis that injuries induced in DNA by UV irradiation switch the copying process since the replica cannot advance further because of the injury in its way. Nevertheless, not even the explanation of the nature of potential injuries (in the case of UV irradiation) and the mechanism of crossing-over could itself guarantee progress toward the understanding of induced recombination. Curtiss (1968), for instance, postulated that the induction effect is connected with the appearance of thymine dimers in DNA, causing local uncoiling of the molecule and thus provoking pairing.

A series of precise experiments undertaken by Howard-Flanders and his collaborators has recently laid the foundations for the elucidation of this problem (Howard-Flanders, 1967; Rupp, 1967; Rupp and Howard-Flanders, 1968a,b; Wilkins and Howard-Flanders, 1968).

On the basis of modern views regarding mechanisms of recombination (4.8) and repair (4.9) it could be supposed that recombination is induced by single-stranded breakages in DNA molecules formed during the excision of thymine dimers. If this hypothesis is correct, recombination in uvr^- strains ought to be induced more effectively than in uvr^+ strains. To verify this hypothesis experiments were carried out on *E. coli.* The recombination frequency in a merozygous system was compared after UV irradiation of uvr^+ and uvr^- (Hfr) donors. Contrary to expectation, the effectiveness of induction was much higher in the second case. This evidently indicated that Curtiss was right and that recombination is in fact induced by unrepaired dimers (the female cell in this experiment was also uvr^-). To verify this conclusion the irradiated Hfr uvr^- chromosome was transferred to a female strain (F^-) capable of repair (uvr^+), in the expectation that the effect of induction would disappear. However, this did not happen.

The solution to the paradox is that recombination is induced, not by dimers, but by other injuries appearing in the Hfr chromosome during replication on transfer to F^-. The nature of these injuries was discovered by the study of a supplementary repair mechanism in uvr^- strains. Although these strains are unable to excise dimers, the mean lethal dose of UV irradiation corresponds to the existence of about 100 dimers in their chromosomes (incidentally, the double mutant uvr^-rec^- dies if there is one dimer in the chromosome). The mechanism for restoring the integrity of the bacterial chromosome in this case is highly original. In the first place, gaps are formed opposite the unexcised dimers in the synthesized DNA strand (Fig. 64). Their formation has been proved by measurement of the sedimentation rate of DNA from UV-irradiated uvr^- bacteria in an "alkaline sucrose" gradient. The decrease in molecular weight of the DNA agreed with that expected

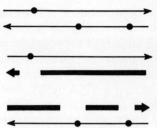

Fig. 64. Diagram illustrating the formation of DNA molecules containing gaps as the result of replication of an unrepaired DNA molecule (after Howard-Flanders, 1967). The thin arrows represent complementary DNA strands, the black circles thymine dimers, and the thick arrows daughter strands.

theoretically. The length of the gaps is of the order of 1000-1600 nucleotides (Iyer and Rupp, 1971). As Fig. 64 shows, after replication of the UV-irradiated DNA two sister molecules are formed, and recombination (analogous to mitotic exchanges between sisters) is induced between them by the presence of the gaps. The recombinant chromosome is formed by utilization of intact segments of the sister chromosomes. Exchanges between sisters have in fact been found in UV-irradiated cells (but not in the control) by equilibrium centrifugation of DNA in a density gradient (Rupp and Howard-Flanders, 1968b; Rupp et al., 1971). Observations (Lehmann, 1972a,b) that the gaps in mammalian cells are filled not with pre-existing, but with newly synthesized DNA are exceptions to this rule.

During conjugation only one DNA strand is transferred to the female cell from Hfr, and the complementary strand is synthesized in the process of the transfer (Rupp and Ihler, 1968; Vapnek and Rupp, 1970, 1971; Wilkins et al., 1971). The transferred chromosome of UV-irradiated Hfr uvr^- appears just as in Fig. 64. In uvr^+ F$^-$ the free ends of the gap are probably protected against the action of exonuclease by the presence of the dimer, whereas the dimer is protected against the action of the reparative system of the F$^-$ cell by the absence of information on the opposite strand. Clearly the recombinant structure containing these double injuries cannot be viable, and accordingly only those recombinants in which short segments of donor fragment, not containing lethals (Shahn, 1968), have been integrated will survive, so that the probability of double exchanges in short segments is increased particularly sharply. This does not mean that only postreplicative induction of recombination can occur. Primary gaps formed after excision of dimers can also induce recombination (Nishioka and Doudney, 1969). Furthermore, Baker and Haynes (1972) clearly showed that recombination is induced in unreplicated chromosomes of phage λ under hcr^- conditions also. In this case the effect may perhaps be due to correctase (see 5.4).

1. It is thus quite evident that recombination is provoked by the presence of single-stranded breakages in the chromosome, in good agreement with modern views on its mechanism (Chapter 5).

2. Second, the potential injuries are injuries of the dimer type leading during replication to the formation of gaps in the synthesized complementary

§4.10] MECHANISM OF INDUCED RECOMBINATION 149

strands, or single-stranded nicks induced by ionizing radiation and radiomimetics and accumulating in the chromosome when DNA synthesis is blocked (Guerola et al., 1971; Unrau and Holliday, 1972). Breakages in one strand are known not to lead to the rapid breakage of the double-helical molecule. They are repaired or they become manifested during replication and recombination. In the presence of regular synapsis these breakages are manifested as recombinations (the reparative function of crossing-over*), but in its absence the result is either nonhomologous crossing-over, leading to chromosomal aberrations, or mitotic recombinations, the frequency of which is determined by the probability of random contacts between homologous segments of the chromosomes.

3. There is a significant difference between nonhomologous crossing-over, leading to structural changes in the chromosomes, and normal crossing-over. The latter always leads to an X configuration of the chromatids, the former equally to U and X configurations, U configurations of chromatids are evidently connected with breakages of strands of different polarities† (Fig. 65). In the same way, in spontaneous aberrations in the circular structures of bacteria either inverted inserts or two rings are formed, depending on the direction of random pairing of the broken strands‡ (Berg and Curtiss, 1967). These facts are a weighty argument in support of the hypothesis that during normal crossing-over the breakages are regulated and take place only in strands of the same polarity in homologous chromatids (5.4).
U exchanges appearing rarely in meiosis are located in places where there are chiasmata (Jones and Brumpton, 1971); they are not observed in mitosis (Couzin and Fox, 1973).

4. The difference in the effectiveness of induction of mitotic crossing-over and conversion is due to the fact that more breakages are required for crossing-over than for conversion (Chapter 5).

---

*Besides the experiments of Schacht and of Howard-Flanders mentioned above, other evidence of the reparative function of crossing-over is given by results obtained by Sparrow (1952). His experiments on *Trillium* showed that the number of fragments counted in metaphase of haploid mitosis of microsporocytes is increased tenfold after irradiation in postpachytene stages of meiosis compared with that observed after irradiation of cells at the pachytene stage. With respect to the two types of reparative recombination in yeasts (postreplicative and "dark," i.e., repairing single-stranded breaks), the papers by Hunnable and Cox (1971) and Fabre (1971, 1972) should be consulted. Multiple reactivation in phages evidently also takes place on account of recombination repair (Rayssiguier, 1972).
†Conclusions regarding the polarity of chromatid subunits, reflecting the polarity of the DNA strands, are drawn from the study of spontaneous (4.3) and x-ray induced (Brewen and Peacock, 1969) exchanges between titium-labeled chromosomes.
‡Nonhomologous reciprocal exchanges of any type, leading to chromosomal aberrations evidently take place at regions of accidental homology. Under these circumstances hybrid genes must be formed and the products of the exchange will be nonviable or mutant. Since not a single mutant was found among hundreds of translocations and inversions, these workers (Berg and Curtiss, 1967) concluded that nonhomologous exchanges take place at the boundaries of the genes (compare Fan, 1969). In any case they do not depend on rec-systems (see the detailed survey by Franklin, 1971a). the frequency of deletions is increased in *polA⁻* (Coukell and Yanofsky, 1971).

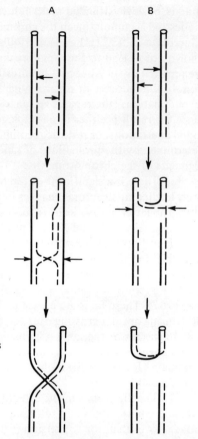

Fig. 65. Diagram showing formation of U and X configurations of chromatids: A) breakages in strands of the same polarity, B) of different polarity. Arrows show sites of breakage. Continuous and broken lines represent DNA strands of opposite polarity.

5. Since the number of repaired injuries is a function of the time elapsing after irradiation and before the beginning of DNA synthesis, the stage of maximal sensitivity to the lethal action of harmful agents coincides with the beginning of the S-period (R. E. Esposito, 1968). The maximal effect toward induction of recombination at this stage can be compared with the effect of induction of recombination in uvr^- strains (Mori and Nakai, 1972). The heterogeneity of the asynchronous cell population in this respect explains the correlation found between induced conversion and crossing-over during mitosis in yeast (Fogel and Hurst, 1963).

6. The suppression of crossing-over in distal regions of *Drosophila* chromosomes by x-rays remains unexplained. All that is clear is that this effect is secondary and is connected with structural differentiation of the chromosome into euchromatin and heterochromatin portions (for a survey and discussion, see Roberts, 1969).

7. Suppression of crossing-over in *Chlamydomonas* by γ-ray irradiation in the preleptotene stage (4.5) likewise has not yet received an unequivocal interpretation. One possible explanation is that small doses of irradiation induce enzymes repairing single-stranded breakages remaining after replication (Lawrence, 1970).

Bridges between strands induced by nitrogen mustard and mitomycin C induce recombination by a mechanism analogous to that which operates in the removal of unexcised dimers (page 148). The only difference is that the first stage is a "bite" produced by *uvrA*, and the diploid state is essential (Cole, 1973).

4.11. Conclusion

1. The chromosomes of eukaryotes and prokaryotes are single-stranded, i.e., they consist of one giant DNA molecule. The structure of this molecule has not been discussed here: it has already found its way into the textbooks.

2. DNA molecules and chromosomes replicate semiconservatively both in mitosis and in meiosis.

3. The copy-choice hypothesis connected genetic recombinations with DNA replication. After the discovery that DNA replication is complete before the cell starts on meiosis it was postulated that pairing of the chromosomes and recombinations take place, not in meiosis, but in the premeiotic stage of DNA synthesis (the S-period). In connection with this hypothesis an extensive series of investigations was undertaken in order to determine at what stages of meiosis these events take place. The results obtained provide convincing evidence that both synapsis and recombinations take place in the prophase of meiosis. This conclusion, together with proof of the semiconservative character of replication of DNA and chromosomes in meiosis, completely rule out any possibility of recombination by replication mechanisms.

4. Recently direct experimental proof has been obtained of the breakage mechanism of crossing-over in both eukaryotes and prokaryotes. A study of recombinant DNA molecules in microorganisms by physicochemical methods demonstrated that recombination takes place in two stages and that limited DNA synthesis, which has also been recorded in meiotic systems, is an integral part of this process.

5. The functional role of this synthesis has been elucidated by the study of genetic control of recombination with the aid of strains of bacteria sensitive to irradiation and of conventionally lethal mutants of bacteriophage T4. A group of different enzymes responsible for breakage, degradation, reunion, and "healing," i.e., reparative synthesis of DNA, has been shown to take part in recombination.

6. As the result of the explanation of these events progress has been made toward the understanding of the mechanisms of spontaneous and induced crossing-over. In the next chapter these mechanisms will be used to explain phenomena discovered during the study of intragenic recombination.

> "A theory is usually the fruit of extreme celerity of the impatient mind, which gladly shuns phenomena and puts in their place forms, concepts, and often even mere words."
>
> *Goethe*

CHAPTER 5

Modern Theories of Genetic Recombination

5.1. Introduction

The direct experimental approach to the study of the mechanism of recombination, which became possible through progress in the development of molecular biology, has proved extremely fruitful (Chapter 4). The problem apparently was solved: it was shown that recombination takes place by breakages and reunions of the DNA molecules and that this process is controlled by appropriate enzymes. Filling in the details could be left to the biochemists. The final stage of work in this direction would be isolation of the corresponding enzymes and performance of the reaction *in vitro*. Without in any way seeking to minimize the necessity, the value, and the prospects of such an approach, the fruitfulness of which was demonstrated by the later work of Anraku and co-workers (Anraku and Lehman, 1969; Anruku et al., 1969) and of Cassuto et al. (1971) it must be pointed out that this would be adequate only if we knew nothing about intragenic recombination. The biochemical concept of recombination in the form in which it was described in Chapter 4 is sufficient to explain crossing-over. However, attempts to interpret phenomena connected with intragenic recombination in terms of this concept have so far proved unsuccessful (5.2).

It was shown earlier in the book (3.8) that interallelic crosses provide information on the processes taking place "inside" chiasmata. Consequently, a general theory of crossing-over could not be formulated without analysis of this information. The theory of crossing-over must therefore be essentially a theory of conversion. Since the converse would be incorrect, for conversion can take place without recombination of the flanks (3.7), this chapter is entitled "Modern Theories of Genetic Recombination" (and not of crossing-over). A theory of genetic recombination must explain the predominant nonreciprocity of intragenic recombinations (3.4), the origin of tetrads containing reciprocal products of recombination between alleles (3.9), the polarity of intragenic recombination (2.11 and 3.6), the allele specificity of conversion (2.12), the map-widening effect (2.7), the interconnection between conversion and crossing-over (3.7), high negative interference (2.9) or, more precisely, conversion without recombination of the flanks (3.2), the genetic control of recombination (3.8), chromosome interference (1.7), and the special features of recombination in prokaryotes (1.7 and 1.18). In this capter the various models of recombination are assessed from the point of view of their ability to explain these facts.

5.2. Random Breakage Models

One of the chief difficulties with which the breakage—reunion hypothesis has had to contend with since it was first put forward is the need to find an explanation of how the precision of the recombination process is achieved down to the molecular level. It must explain the fact that recombinant mole-

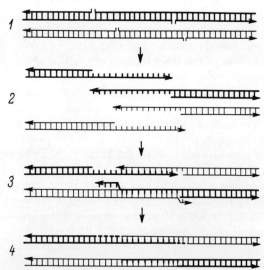

Fig. 66. The model of recombination proposed by Thomas (1966): 1) the two recombining DNA molecules (breakages in the complementary strands do not lie opposite one another); 2) splitting of parental molecules into fragments with free single-stranded ends; 3) reunion of fragments in pairs, leading to the formation of joint molecules; 4) repair of the gaps and excision of excess material; conversion of the joint molecules into recombinant.

cules contain neither more nor less genetic material than the parental molecules. If recombining double-helical molecules are broken at precisely homologous points, it is not clear by what mechanism the double-helical fragments are arranged end to end as is required for the formation of phosphate bonds. It was accordingly postulated that breakages in the parental molecules during recombination do not take place at strictly homologous points, but a short distance apart, so that reunion of the fragments becomes possible through the formation of hydrogen bonds between the complementary strands in the region of partial overlapping (Fig. 60) (Levinthal, 1954).*

Before such an association can take place, the ends of the fragments must be partially uncoiled (terminal dissociation). According to Szybalski (1964), after breakage of the parental molecules their free 3'-ends are attacked by exonucleases, exposing the ends of the overlapping fragments (Fig. 63). Single-stranded ends can also arise if breakages in the complementary strands of the two molecules do not coincide (Fig. 66). The result depends on the mutual arrangement of the breakages (Thomas, 1966). In the chromosomes shown in Fig. 66 the breakages are so arranged that during reunion of the fragments two joint molecules are formed, one containing gaps, the other an excess of DNA. Nevertheless such molecules can easily be converted into recombinant by exonucleotic removal of the excesses and reparative replication of the gaps. If, however, breakages in the complementary strands of one molecule in this figure change places, the reciprocal molecule cannot be formed. Calculations based on the hypothesis of the random character of distribution of the breakages shows that the efficiency of formation of reciprocal recombinants in the pool of vegetative phage does not exceed 50% (Thomas, 1966).

Thomas considers that interaction between the fragments is random in character, since the homologous chromosomes of prokaryotes cannot be prearranged relative to each other in a state analogous to the synapsis of eukaryote chromosomes. In fact, we do not yet know what forces could be responsible for association of double-helical DNA molecules other than those which arise between complementary bases (see Mosevitskii, 1969).

Since breakages and interaction of the fragments during recombination are random in character, high negative interference cannot be explained by the hypothesis of effective pairing. It can be explained (Watson et al., 1966; Doermann and Parma, 1967) on the assumption that individual single-stranded fragments of parental chromosomes are fitted into the unrepaired gaps in the joint molecules (Fig. 67). Filling in the gaps in this way clearly simulates double exchange, and if the number of events of this type is large

---

*Unwinding and pairing can, in principle, precede primary breaking. Spontaneous denaturation is known to take place at a physiological temperature in A–T ridge sites (page 142) within the molecule, leaving the molecule as a whole intact. The authors of several models of genetic recombination emphasize this point (Fogel and Hurst, 1967; Sermonti and Carere, 1968; Moore, 1972; Sobell, 1972).

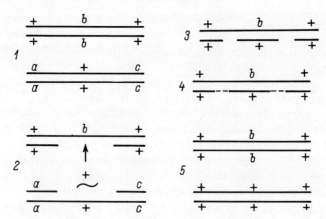

Fig. 67. Model of double crossing-over (after Doermann and Parma, 1967): 1) parental DNA molecules; 2, 3) single-stranded fragment of one parental molecule is fitted into (marked by the arrow) an unrepaired gap in the homologous molecule; 4) repair; 5) replication of the heterozygous molecule leading to the formation of a wild-type recombinant.

enough and the size of the "traveling" fragments small enough, from the formal point of view this explanation appears satisfactory.

The possibility of conversion of the gene follows naturally from the mechanism of random crossing-over, for in this case there is partial loss and resynthesis of DNA in the region of union of the fragments – "exonucleotic repair" (Taylor et al., 1962; Whitehouse, 1963; Meselson, 1965; Mosevitskii, 1968; Stahl, 1969).

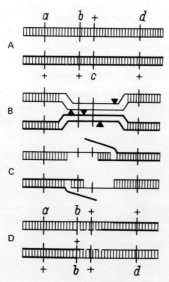

Fig. 68. Model of gene conversion (from Mosevitskii, 1968). Only two recombining chromatids are shown. A) Original state, B) random breakages in the region of effective pairing, C) union of the fragments, D) exonucleotic repair leading to conversion $c \rightarrow +$ and to aberrant segregation 4+:4− with respect to the b marker.

The character of recombination of the alleles in the region of reunion is determined by the arrangement of the breakages. This arrangement in Fig. 68 is such that the marker c undergoes conversion, marker b is in a heterozygous state, and markers a and d, which are outside the breakage zone, recombine normally. Although the possibility of exonucleotic conversion has been demonstrated, in principle, *in vitro* (Bautz et al., 1968), this idea has only psychological value, for it shows that an explanation of conversion can be found within the framework of the breakage–reunion mechanism without the need to invoke the copy-choice hypothesis. The attempt to explain the facts summarized in Section 5.1 on this basis has been unsuccessful. Conversion as a result of exonucleotic repair is invariably connected with crossing-over (a single mechanism). Evidence of the existence of two different mechanisms of genetic recombination was presented in Section 3.7. Furthermore, as Fig. 68 shows, the arrangement of the breakages is often such that reciprocal recombinants are formed (a and d do not convert). It was shown in Section 3.4 that conversion is the main method of recombination between alleles. Moreover, the appearance of t octads is allele-specific. It is completely unknown how the allele can determine the mutual arrangement of the breakages. The polarity of recombination likewise remains inexplicable. A characteristic feature of Thomas's model is that one of the chromosomes taking part in recombination is not preserved in half of the cases but breaks up into fragments. In crossing-over, however, preservation of both recombinant chromosomes must occur in 100% of meioses. For this purpose, breakages in individual strands must be fixed relative to each other or on the chromosome. Crossing-over must be a controllable process.

5.3. Taylor's Model

The discovery of the phenomenon of polarization of conversions in the gene led to acceptance of the idea that the organization of the chromosome is discontinuous with regard to recombination events (2.11 and 3.6). This idea has been further developed in all recent models of recombination.

Holliday (1964b) and Hastings and Whitehouse (1964) independently postulated more or less simultaneously that primary breakages are fixed on the chromosome at points corresponding to special bonds. Since there is no evidence of this in the chromosome (4.2), it is suggested that the recombination discontinuity of the chromosome is determined by specific base sequences analogous to the operator or promotor regions. Holliday (1968) calls this group of bases the recombiner, and Whitehouse (1966) the operator. Taylor (1967) postulated that polarons can be identified with replicons and that some of them embark upon meiosis in an unreplicated form (compare with Stadler's model; 4.7). After cytological synapsis there is association between the strands of opposite polarities (Fig. 69), the primary breakages taking place at opposite ends of the homologous replicons. Events then

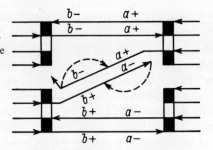

Fig. 69. The model of recombination suggested by Taylor (1967). Arrows show the direction of phosphate bonds. The black squares are boundaries of replicons starting meiosis in a partially unreplicated form. Broken arrows show the direction of exonucleotic degradation of DNA.

take place exactly as in the system *in vitro* (5.2): exonuclease III destroys one of the complementary strand as template. If a site of heterozygosity is present in the replicon conversion takes place. The probability of conversion is determined by the position of the site in the replicon. It is greatest at the borders and least in the center. This model, like the "sex circle" model (Stahl, 1969) and Paszewski's (1970) model, suffers from the same defects as the models of Stadler and Freese (for critical comments, see 2.11, 3.6, 4.7, and also Holliday and Whitehouse, 1970).

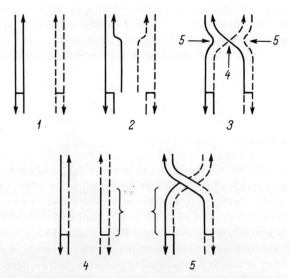

Fig. 70. The model of recombination suggested by Holliday (1964b): 1) homologous chromatids (the cross lines represent the recombiner); 2) dissociation of strands of identical polarity in homologous recombiners; 3) secondary breakages, numbers by arrows indicate result (4 or 5) of breakages at corresponding points; 4) noncrossover (P) configuration with hybrid DNA in identical sites of both homologous chromatids (the hybrid region if indicated by brackets); 5) crossover (R_1) configuration.

5.4. Holliday's Model (Holliday, 1962, 1964b, 1966a, 1968)

Polarity. Discontinuity of the chromosome with respect to recombination events is produced by recombiners in which primary breakages take place (Fig. 70: 1, 2). Holliday postulates that DNA strands of identical polarity are broken in homologous chromosomes. The primary breakages are evidently produced by specific endonucleases. The hypothesis that endonuclease can act selectively on a certain sequence of nucleotides seems at present to be very well founded, in view of the accumulating information on regulatory proteins and enzymes possessing this type of specificity (the repressor: Ptashne, 1967; Adler et al., 1972; limiting and modifying enzymes: Arber, 1964, 1971; Kelly and Smith, 1970; Hedgpeth et al., 1972; Haberman et al., 1972; Smith and Wilcox, 1970; the integrases* of bacteriophages $\phi 80$: Signer and Beckwith, 1966; λ: Echols et al., 1968; Weil and Signer, 1968; Gottesman and Weisberg, 1971; P2: Lindahl, 1969).

The primary "nicks" initiate longitudinal dissociation of the individual DNA strands (Fig. 70: 2). This process has been shown to be facilitated by sorption of the corresponding protein. Specific association of the complementary DNA strands arising from different homologues next takes place (Fig. 70: 3). After crossed pairing (association) a configuration which has been called the semichromatid chiasma is formed.

Secondary breakages take place either in the same strands as the primary nicks (i.e., in strands forming a half-chiasma), or in intact strands (Fig. 70: 4, 5). In the latter case the half-chiasma is converted into a complete chiasma (crossing-over). Secondary breakages and reunions take place with high accuracy, so that neither deletions nor duplications are formed in any of the recombinant chromatids, and there is neither degradation nor resynthesis of DNA. Whatever the case, alleles on both sides of the original half-chiasma recombine. If, therefore, the basis of crossing-over is chiasma formation, the basis of recombination in short segments of genetic material (within the gene) is the mechanism of formation of half-chiasmata. I have ventured to call this last type of genetic recombination "semicrossing-over." The two recombinant chromatids thus formed consist of parental strands of DNA in segments terminal to the recombiner and of hybrid† segments at the proximal ends. The closer the allele is to the recombiner, the more probably

---
*Integration crossing over takes place in the region of something of the order of 12 base pairs and it is not based on homology $att\phi$ and $att\beta$. Vegetative recombination under the influence of the Int system is always reciprocal (1.18) and it also involves a very limited region of the chromosome, as a result of which the recombinants form a sharp peak in a density gradient (Nash and Robertson, 1971). Integration crossing over is not accompanied by additional DNA synthesis. The efficiency of transfer of the parental label to the recombinants may reach 50% of more (Kellenberger-Gujer and Weisberg, 1971). Parkinson (1971) showed that there is definite asymmetry and irreversibility of the recombination events taking place under the influence of the Int system.

†The term "hybrid" is taken to mean a DNA molecule whose complementary strands are obtained from different parents (heteroduplex).

it lies in a hybrid region. In a one-point ($a \times +$) cross a site of heterozygosity arises in this region, and acts as a trigger for conversion (see below). A gradient of hybridization (and, consequently, of conversion also) thus is established in the region of the recombiner.

In Holliday's opinion, recombiners are distributed irregularly throughout the chromosome and they are not directly related to the boundaries of the genes. Recombiners, moreover, can be determined by base sequences coding amino acids, and for that reason they are more likely to be found within genes than between them. To explain polarity it must be accepted that the number of recombiners is small and that usually the nearest recombiner lies outside the gene in which recombination is being studied. The suggested mechanism rules out the possibility of using exonucleotic repair to explain conversion: if the free ends of the strands of the recombiners (Fig. 70) are 3'-ends, exonucleotic repair simply restores the original state (Fig. 70: 1). The absence of any subsequent events leads to the appearance of a 4+:4−ascus with an aberrant spore arrangement (reciprocal semichromatid conversion; 3.9). Consequently, wild-type recombinants will appear as 1+:7−asci in two-point crosses. Since asci of this type are found only exceptionally, further ideas are required.

Correction of Molecular Heterozygosity. Holliday postulated that sites of heterozygosity are corrected by replacement of one of the unpaired bases by a suitable free base (Fig. 71). At that time (1962) nothing was yet known about the mechanism of dark repair. In the modern view a more realistic mechanism is that which has been called (Meselson, 1965) endonucleotic repair. The hypothesis that correction of molecular heterozygosity can be corrected by a mechanism of this type was expressed by several people independently (Dulbecco, 1964; Pritchard, 1964, cited by Holliday, 1968; Whitehouse and Hastings, 1965).

If hybrid DNA is formed in both recombining chromatids, depending on the direction of correction the following outcomes are possible (Fig. 72). Correction in the same direction (i.e., either toward + or toward −) in both chromatids (A and B) leads to the conversions $a \rightarrow +$ and $+ \rightarrow a$ (3+ : 1− and 1+ : 3−). Correction in different directions (C and D) leads either to restoration of the original state or to the formation of a t tetrad (reciprocal conver-

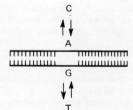

Fig. 71. Diagram of correction of a site of heterozygosity in the DNA molecule (after Holliday, 1962). A, T, G, and C represent bases.

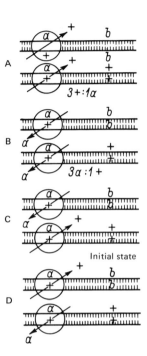

Fig. 72. Diagram showing formation of different types of asci as the result of correction. In each variant two of the four chromatids are shown. Hybrid DNA occupies site a in both recombining chromatids. A and B) Correction in both chromatids in one direction, C and D) in different directions. Sites of heterozygosity are circled. Arrows indicate direction of correction (compare with Fig. 76).

sion). If correction in either direction takes place in only one chromatid and the other remains heterozygous, postmeiotic segregation is observed (types of asci 5+ : 3− and 5− : 3+). It thus follows that the name "semichromatid conversion" does not accurately reflect the true state of affairs. If correction took place in none of the chromatids, 4+ : 4− asci with an aberrant arrangement of the spores would be formed. On the assumption that hybrid DNA is formed in both recombining chromatids at a given site, the probability of appearance of any class of asci can be calculated (Whitehouse and Hastings, 1965). If the probability of correction toward + and toward − is designated p^+ and p^- respectively, and the probability of absence of correction p^0, we have $p^+ + p^- + p^0 = 1$. The expected frequencies of appearance of 4+ : 4− asci with an aberrant arrangement of the spores, and of 5+ : 3−, 3+ : 5−, 6+ : 2−, and 2+ : 6− spores in that case will be given by $(p^0)^2$, $2(p^0 p^+)$, $2(p^0 p^-)$, $(p^+)^2$, and $(p^-)^2$ respectively (Table 11). It is clear that the relative values of p^+, p^-, and p^0 can vary widely depending on the type of mutation injury (allele specificity).

Analysis of results obtained in their experiments by Kitani et al. (1962) shows that values of p^+, p^-, and p^0 can be chosen by the method of successive approximations so that the mathematical expectancies of the different types of asci calculated from them agree closely with the experimental data (column I in Table 12, from Whitehouse and Hastings, 1965).

TABLE 11. Expected Frequencies of Appearance of Different Types of Asci if Hybrid DNA Is Formed in Both Chromatids (after Whitehouse and Hastings, 1965)

Parameters of correction	p^+	p^-	p^0
p^+	6+:2— $(p^+)^2$	4+:4— p^+p^-	5+:3— p^+p^0
p^-	4+:4— p^-p^+	2+:6— $(p^-)^2$	3+:5— p^-p^0
p^0	5+:3— p^0p^+	3+:5— p^0p^-	(4+:4—)$_a$ * $(p^0)^2$

*Ascus with an aberrant arrangement of the spores.

The theoretically expected values in column II of this table are calculated on a different basis (Emerson, 1966a). Emerson postulated that the relative values of p^+, p^-, and p^0 for each particular allele may differ in different chromatids. This may evidently happen because the character of disturbance of complementarity in the site of heterozygosity must be different in these chromatids if the initial breakages take place in strands of equal polarity.

In fact, if a simple replacement

$$\begin{array}{cc} \uparrow & \uparrow \\ A-T \rightarrow & G-C \\ \downarrow & \downarrow \end{array}$$

TABLE 12. Comparison of Mathematical Expectancies of Different Types of Asci with Aberrant Segregation for Gray (Spore Color) with the Experimental Data (Kitani et al., 1962) Obtained by Crossing the Mutant (Gray) with the Wild Type (+)

Type of ascus	Number of asci		
	Expected theoretically		Found
	I *	II	
6+:2—	94.5	98.3	98
2+:6—	10.5	13.0	13
5+:3—	113.5	107.8	108
3+:5—	23.0	20.0	20
(4+:4—)$_a$	12.5	4.5	9

*Values of parameters: $p^0 = 0.27$; $p^+ = 0.56$, and $p^- = 0.17$.

took place as the result of mutation, the sites of heterozygosity in the recombinant chromatids would be

$$\overset{\uparrow}{\underset{\downarrow}{\text{A--C}}} \text{ and } \overset{\uparrow}{\underset{\downarrow}{\text{T--G}}}$$

(the arrows denote the polarity of the complementary strands). Let n be the total number of corrections in chromatid 1 and r the fraction of correction events (of n) in chromatid 1 to the wild type (+). In that case nr is the fraction of correction to +; $n(1-r)$ the fraction of correction to the mutant state (−), and $(1-n)$ the fraction of absence of repair. Let q and s correspond to the values of n and r for chromatid 2. In Table 13 opposite each type of ascus is shown the corresponding probability of its formation calculated on this basis. The last column of the table gives the same experimental data as in Table 12, but expressed in frequencies of appearance of asci of a particular type (Kitani et al., 1962).

Comparison of the observed frequencies of different types of asci with the theoretical probabilities of formation of each type suggests that the two sites of heterozygosity do in fact differ from each other in the frequency of correction and that they differ even more in the relative frequencies at which this correction takes place toward the wild type than toward the mutant.

From the relative frequencies of 5+ : 3− and 3+ : 5− asci

$$\frac{n(1-q)r + (1-n)qs}{n(1-q)(1-r) + (1-n)q(1-s)} = \frac{108}{20},$$

$$n = \frac{q(325-27)}{q(32r + 32s - 54) + 27 - 32r}.$$

(1)

TABLE 13. Types of Asci and Probability of Their Formation (after Emerson, 1966a)

Event in chromatid 1	Event in chromatid 2	Type of ascus	Probability of appearance of wild-type asci	Observed frequency of appearance of asci of corresponding type
c → +	c → +	3+ : 1−	$nqrs$	98/239
c → +	c → −	2+ : 2−	$nqr(1-s)$	?
c → −	c → +	2+ : 2+	$nq(1-r)s$	
c → −	c → −	1+ : 3−	$nq(1-r)(1-s)$	13/239
c → +	nil = +/−	5+ : 3−	$n(1-q)r$	108/239
nil = +/−	c → +	5+ : 3−	$(1-n)qs$	
nil = +/−	c → −	3+ : 5−	$(1-n)q(1-s)$	20/239
c → −	nil = +/−	3+ : 5−	$n(1-q)(1-r)$	
nil = +/−	nil = +/−	(4+ : 4−)$_a$	$(1-n)(1-q)$	9/150

Note. c) Correction, nil) absence of correction.

From the relative frequencies of 5+ : 3− and 6+ : 3− asci

$$\frac{n(1-q)r + (1-n)qs}{nqrs} = \frac{108}{98},$$

$$n = \frac{q(49s)}{q(49r + 49s + 54rs) - 49r}.$$
(2)

Dividing (1) by (2) we obtain

$$q = \frac{s-r}{s-r-1.102rs + 1.306rs^2}.$$
(3)

From the relative frequencies of 6+ : 2− and 2+ : 6− asci

$$\frac{nqrs}{nq(1-r)(1-s)} = \frac{98}{13},$$

$$r = \frac{98 - 98s}{98 - 85s}.$$
(4)

Consequently, for each arbitrary value of s numerical values of r [by Equation (4)], q [by substituting s for r in Equation (3)], and n [by substituting s for r and q in Equations (1) or (2)] can be obtained. The probable values of all three parameters as functions of the fourth (s) are shown in Fig. 73. Choice of the values of these parameters corresponding to the greatest probability of appearance of asci of the (4+ : 4−)$_a$ type gives surprisingly accurate agreement between the theoretically expected probabilities and the experimental data (Table 14 and column 2 of Table 12).

High Negative Interference (Interpretation of Distribution of Flank Markers). In the standard flank system wild-type recombinants are formed as a result of conversions of a or b alleles. Let us exclude t tetrads from the analysis. We are left to interpret eight flank classes (Table 5). If dissociation begins at the proximal end, two principal flank classes, P_1 and R_1, can be formed because conversion takes place at site a (Fig. 74). If the half-chiasma

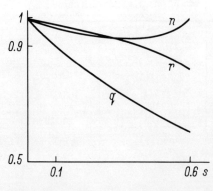

Fig. 73. Determination of the parameters n, r, q, and s for the g_1 mutant of *Sordaria fimicola* (after Emerson, 1966a). Abscissa, values of parameter s; ordinate, values of n, r, and q as fractions of s.

TABLE 14. Values of Parameter p Selected for Asci with Aberrant Segregation with Respect to g_1 Based on the Assumption of Equal Effectiveness of Correction in Different Chromatids

Emerson's parameters	Whitehouse's parameters (Table 11)		
	p^+	p^-	p^0
$n = 0.935$ $r = 0.92$	0.86	0.07	0.07
$q = 0.71$ $s = 0.4$	0.28	0.42	0.3

is not converted into a complete chiasma, the P_1 configuration is formed; if it does, the R_1 configuration is formed. Furthermore, if dissociation begins at the distal end, P_2 and R_1 configurations are formed. The R_1 class thus appears whatever the type of dissociation. The ratio between the P_1 and P_2 classes is determined by the frequency of dissociation at different ends of the gene. Since in all systems investigated both P classes appear, the genes will never be strictly unipolar, but predominant polarity (predominance of one P class over the other) is found in many places (it is least clearly ex-

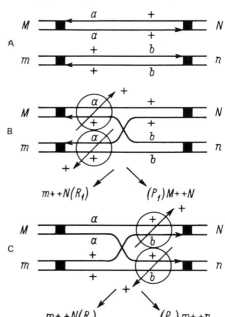

Fig. 74. Mechanism of formation of P and R_1 classes in a standard flank system (after Holliday, 1968). A) Initial state, B) dissociation in the proximal recombiner, C) in the distal. Only two of the four chromatids are shown. Black squares represent recombiners. Remainder of legend as in Fig. 72.

pressed in *Neurospora crassa,* see the survey: Catcheside, 1966b). The P configurations in the standard flank system are thus explained by Holliday's hypothesis by conversion in the half-chiasma, i.e., without complete (chromatid) crossing-over. Consequently, there is no need to postulate, as Pritchard did, high negative interference and multiple crossovers in the region of effective pairing. The P_1 and P_2 classes, which can be interpreted as products of double exchanges, are in fact single conversions without any accompanying crossing-over (without the change from the half to the complete chiasma), while the triple exchange (R_2 class) is simulated by conversion with simultaneous crossing-over on the other side of the converted allele (Fig. 75), taking place at normal frequency (for a discussion on the absence of interference between conversion and crossing-over, see 3.7). The $P_2 c(a \rightarrow +)$ and $P_1 c(b \rightarrow +)$ classes evidently arise from the corresponding R_1 classes as the result of additional crossing-over on one or other side of the gene: $P_2 c(a \rightarrow +)$ from $R_1 c(a \rightarrow +)$ as the result of crossing-over in the distal region and $P_1 c(b \rightarrow +)$ from $R_1 c(b \rightarrow +)$ as the result of crossing-over in the proximal region. All classes (R and P) with conversion of the proximal a allele are thus formed as the result of dissociation at the proximal end of the gene, while all R and P classes with conversion of the b allele are formed as the result of a distal process of dissociation.

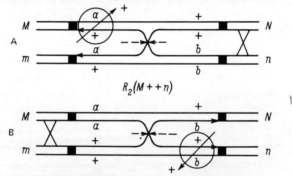

Fig. 75. Mechanism of formation of R_2 class in a standard flank system (After: Holliday, 1968). A) Dissociation in the proximal recombiner. Formation of hybrid DNA ends between sites a and b; the site of heterozygosity $a/+$ in chromatid MN is corrected toward the wild type. The P_1 configuration formed is converted into the R_2 configuration as the result of additional crossing-over in the distal segment. B) Dissociation in the distal recombiner, conversion in site $+/b$, accompanied by crossing-over in the proximal region, also leads to the formation of an R_2 configuration of the flank markers. Broken arrows indicate sites of secondary breakages. Oblique crosses denote additional exchanges. Remainder of legend as in Figs. 72 and 74.

If the suggested interpretation of the formation of flank classes is correct, the relative frequencies of formation of hybrid DNA at both ends of the gene can be determined. Furthermore, if the probability of conversion of the half-chiasma into a whole chiasma is independent of the end of the gene at which dissociation takes place, the relative frequencies of the dissociations can be determined in two ways: by comparing classes formed in half-chiasmata and classes formed in whole chiasmata. Classes P_1c ($a \to +$) and R_2c ($a \to +$) are formed from a half-chiasma with proximal dissociation. Classes P_2c ($b \to +$) and R_2c ($b \to +$) are formed from a half-chiasma with distal dissociation. Consequently, the ratios

$$\frac{P_1c\ (a \to +)}{P_2c\ (b \to +)} \tag{1}$$

and

$$\frac{R_2c\ (a \to +)}{R_2c\ (b \to +)} \tag{2}$$

show by how many times the frequency of dissociation is higher at one end of the gene than at the other. Substituting the frequencies of the corresponding classes from Table 6 we obtain that in the *hi-1* gene the dissociation cycle is initiated either 8 times (from ratio 1) or 20 times (from ratio 2) more frequently at the proximal end of the gene than at its distal end. In the same way, for classes formed in whole chiasmata,

$$\frac{R_1c\ (a \to +)}{R_1c\ (b \to +)} \tag{3}$$

and

$$\frac{P_2c\ (a \to +)}{P_1c\ (b \to +)}. \tag{4}$$

It follows from these ratios (data from Table 6) that the frequency of proximal dissociation is 3.5 times (from ratio 3) or more than 7 times (from ratio 4) greater than that of distal dissociation. The slight discrepancy between the results is due to the fact that the difference between the length of the flanges must be allowed for in ratio (2) and interference in ratio (4). The only important point which remains unexplained by Holliday's model is the difference between the ratios (1) and (3).

Nature of Reciprocal Intragenic Recombination. The main difficulty facing these correction models, and it has not yet been overcome, is that they cannot explain the predominant nonreciprocity of intragenic recombination. As Table 11 shows, if $p^- = p^+$, tetrads containing reciprocal products must appear in about 50% of cases. As we have seen (3.4) this situation arises only exceptionally. An excess of c tetrads may be the result of marked inequality in the basic frequencies of conversions of the allele concerned (if

$fk_{(a \to +)} \geqslant fk_{(+ \to a)})$, but sharp differences are found only as exceptions and they do not correlate with the increased frequency of appearance of t tetrads in crosses involving these alleles. Another explanation which has been suggested is that hybrid DNA is formed in only one chromatid. In that case the formation of reciprocal recombinants would be completely impossible (Fig. 76).

Holliday at first tried to escape from the predicament with the t tetrads by postulating that heterozygosity leads to unstable association of the strands, so that only a site in one chromatid can be corrected. Crossed pairing then ceases and the threads revert to their original partners. However, a half-chromatid chiasma is still formed, for the free ends of the hitherto

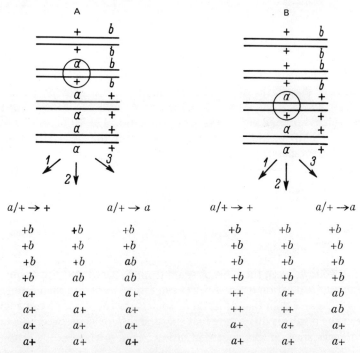

Fig. 76. Possible types of asci in the case of formation of a site of heterozygosity in only one of the recombining chromatids: A) formation of a site of heterozygosity through excessive copying of the a allele (compare with Fig. 77, 4 or with Fig. 78); B) excess copying of the + allele (compare with Fig. 77, 3). 1) Correction toward the wild type; 2) absence of correction; 3) correction toward a. Genotypes of these spores are shown below.

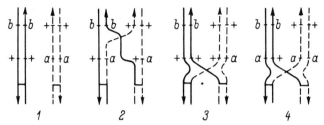

Fig. 77. Mechanism of unstable association suggested by Holliday to explain the nonreciprocity of intragenic recombination (after Holliday, 1964b). The possibility of conversion is limited to interference of the site of heterozygosity $a/+$ with the formation of hybrid DNA. 1) Original state; 2) unstable association in the right chromatid; after correction of the site of heterozygosity in the left chromatid a half-chromatid chiasma is formed between site a and the recombiner; 3) correction in the left chromatid at the second stage toward the wild type; 4) correction at the second stage toward a.

associated strands have now become attached to the recombiners (Fig. 77).* The basis of this idea is the assumed inhibition of conjugation (see page 215). To explain the origin of the t tetrads Holliday (1968) put forward the hypothesis of "true" crossing-over between alleles.† Since the recombiner never lies within the gene studied, the formation of hybrid DNA and "true" crossing-over can take place between mutant sites without affecting them, so that reciprocal recombination results. However, since the recombiner is fixed to the chromosome, allele specificity of reciprocal recombination still remains completely unexplained. Furthermore, whereas in certain cases this idea can be used to interpret recombination in complex loci (see Touré, 1972b), it provides no basis for the explanation of predominant nonreciprocity. If, for example, the recombiner lies in the *hi-1* gene (Table 6), as Holli-

---

*Many workers have emphasized that the situation illustrated in Fig. 76 must be taken as the basis for the explanation of nonreciprocity, regardless of whether they assume a role of the correction system or not. The situation may be the result of unequal participation of the parents in the recombination, connected with transformation-like transfer of material from the donor chromatid to the recipient chromatid. See the "poisoned arrow" model of Hotchkiss (1971), the model of Stadler and Towe (1971), and also the models of Paszewski (1970) and of Boon and Zinder (1969, 1971). The same situation can result from the mechanism of strand assimilation proposed by Cassuto and Radding (1971) on the assumption that the gaps in the donor chromatid are repaired synchronously by DNA-polymerase. However, all these explanations are *ad hoc*. For example, generalization of the strand assimilation mechanism leads to types of models with which we are already familiar (5.2, 5.3). The foregoing remarks do not rule out the possibility that conversion is effectively induced by fission of the P^{32} incorporated in only one gene of the yeast zygote (Korolev and Gracheva, 1972).

†In "true" crossing over between alleles the hybrid DNA lies in the region between the mutant sites without affecting either of them, so that the possibility of conversion is ruled out (see also page 173).

day supposes, it is not clear why t tetrads are formed in a low frequency (below 10%) in this system.

If as the result of a mutation event a new (additional) recombiner appears in the middle of a gene (as may happen in the case of mutation *M26* in yeast (Gutz, 1971b,c) this leads to an increase in only the c and not the t tetrads.

5.5. The Whitehouse–Hastings Model (Whitehouse, 1963, 1966, 1967a; Hastings and Whitehouse, 1964; Whitehouse and Hastings, 1965)

Polarity. Primary breakages take place in operators. [They perhaps remain unrepaired after the premeiotic S-phase (Hastings, 1972; Whitehouse, 1972).] In homologous chromatids (Fig. 78: 1) strands of different polarity are broken. A process of dissociation (Fig. 78: 2) is initiated from the points of breakage in one direction and DNA synthesis takes place along the intact strands (Fig. 78: 3). The crossover molecules arise by pairing between old and new polynucleotide chains (Fig. 78: 4-6). Breakage and degradation of the parental strands still left intact complete the process of formation of recombinant chromosomes (Fig. 78: 7, 8). Events taking place during recombination recall the events which must take place during transcription.

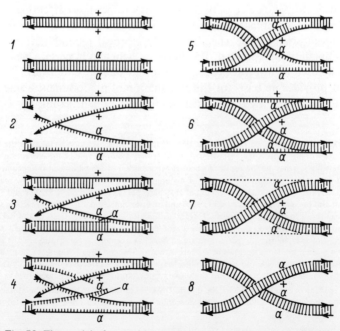

Fig. 78. The model of recombination suggested by Whitehouse (1965).

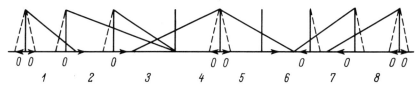

Fig. 79. Diagram illustrating the possible origin of different types of polarity in genes on the basis of the operator model of crossing-over (after Whitehouse, 1966). The horizontal line represents a segment of chromosome with eight genes. 0, operator. Arrows indicate the direction of dissociation (polarity). Sloping lines represent the gradients of formation of hybrid DNA. Broken lines represent hybrid DNA of reverse crossing-over (compare with Fig. 82).

Additional DNA synthesis, starting in the operator, simulates transcription, the only difference being that this synthesis "dies out" toward the terminal part of the gene. Accordingly the process of formation of hybrid DNA is expressed as a gradient: hybrid DNA is formed most frequently in the operator end of the gene and least frequently in the terminal end. Since the basic frequency of site conversion (fk) is determined by the possibility of incorporation of the site into the hybrid region, the observed gradient fk reflects the gradient of formation of hybrid DNA in the gene.

Analysis of the published data on types of polarity led the authors of this hypothesis to conclude that unipolar and bipolar genes must exist, and in the bipolar type the relative frequencies of formation of hybrid DNA from different ends of the gene must vary arbitrarily. To fit these conclusions into the operator hypothesis (the presence of two operators in a gene is, to say the least, doubtful), Whitehouse postulated that both in recombination and in transcription dissociation of the DNA strands begins at only one (0) end of the gene (the operator), but that dissociation and the formation of hybrid DNA connected with it spread throughout one gene and into the next.* The interlinking of neighboring genes by the single dissociation process means that different types of polarity can exist. The type of polarity in a gene is determined by the direction of dissociation in the neighboring gene next to the end of the first gene (the T end) removed from the operator. If the direction of dissociation in this gene coincides with its direction in the first gene, unipolarity will be exhibited (Fig. 79: 1, 2, 7, and 8). If the directions of dissociation in the neighboring genes are different (i.e., the genes are in contact with each other by their T ends), both genes will be bipolar (3, 4, 5, 6). If the length of the hybrid DNA formed at the 0 end is approximately constant, the type of polarity in the gene will depend to some extent on its length.

The operator hypothesis postulates the identity of the site specific for the repressor and the site in which dissociation is initiated during recombi-

---

*How far nobody knows. In some cases, however, it must be assumed that this process extends over as many as $8 \times 10^5 - 9 \times 10^5$ base pairs.

nation. In the *hi-1* gene the character of polarity observed (predominantly proximal) agrees with this hypothesis, for the operator is known to be localized in the proximal end of the gene (Fogel and Hurst, 1967). However, data on the connection between transcription and intragenic recombination obtained in work on different systems are highly contradictory and have not yet lent themselves to unequivocal interpretation (Clavilier et al., 1960; Helling, 1967; Shestakov and Barbour, 1967; Herman, 1968a; Cooper and Fox, 1969; Savić, 1972; Savić and Kanazir, 1972). In addition, the model has to contend with an obvious difficulty (Holliday, 1968), for the points of primary breakages must be on strands of different polarity in homologous chromatids, whereas the repressor must exhibit specificity only toward strands of the same polarity. As Emerson (1966a) showed, the equality of the p parameters which follows from Whitehouse's scheme in many cases does not exist (Table 14). This is also a powerful argument in support of the view that different pairs of noncomplementing nucleotides arise in the chromatids, as Holliday's scheme requires.

Gutz (1971a) observed that equality of the p parameters is not initiated in Whitehouse's model, for the type of mispairing will vary depending on which strands (right or left) are broken. In this case, however, the types of mispairing will differ in different meioses, although as before they remain the same in the same meiosis.

Origin of the t Tetrads. DNA replication was introduced into Whitehouse's model primarily because it could provide the basis for an explanation of the predominant nonreciprocity of intragenic recombination (Whitehouse, 1970). Unlike Holliday's model, in which hybrid DNA must be present in both chromatids, in the operator model this difficulty is removed by the assumption that DNA synthesis in chromatids can be unequal, so that the hybrid DNA in one of them may be very short and may not reach the mutant site a (Fig. 78). Whitehouse and Hastings (1965) introduced the term "strand coefficient," which is determined as follows:

$$\frac{t + 2c\,(ab \to +)}{c\,(a \to +) + c\,(b \to +)}.$$

Events of the $c(ab \to +)$ type in *trans*-two-point* crosses are very rare (see page 189), so that the formula can be written in the form

$$\frac{t}{c\,(a \to +) + c\,(b \to +)}.$$

Since it is understood that t tetrads arise only when the hybrid DNA is in a mutant site in both chromatids, the strand coefficient shows the relative frequency of its appearance. If the strand coefficient is 0, no t tetrads what-

---

*Alleles taking part in the cross lie on different chromosomes $(a \times b)$.

ever are formed, for the hybrid DNA never reaches the mutant site in both chromatids simultaneously. This situation is characteristic of *46* series of *Ascobolus immersus* (3.4). If t tetrads are observed to appear, the strand coefficient has an intermediate value between 0 and 1.

Two observations conflict with the hypothesis of origin of the t tetrads described above. The first is the allele specificity of reciprocal recombination. There is absolutely no explanation how the character of mutation damage at a site hundreds of nucleotides away from the ends of the gene can influence the rate of DNA replication initiated in the operator, or the length of its replication in either chromatid. The second is that in the overwhelming majority of cases the t tetrads have an R_1 configuration of the flanks (3.6). At first glance it seems evident that if t tetrads appeared as the result of reciprocal conversion, the distribution of the flanks in them would be the same as in c tetrads.

From these considerations it can be postulated (Fogel and Hurst, 1967; Whitehouse, 1967a) that t tetrads appear, not as the result of reciprocal conversion, but through "true" crossing-over between alleles (in "true" crossing-over hybrid DNA is present only in the region between the mutant sites).

What is the character of the hypothetical events in this case? If t tetrads appeared in all crosses it might be supposed that, besides regular recombination events starting in the operators, crossing-over with a random arrangement of breakage points can also occur within the gene. However, the allele specificity of reciprocal recombination prevents this explanation. Moreover, there is reason to suppose that during the formation of t tetrads recombination is initiated in the usual way from one end of the gene. In a *277 X 63* cross (series *46*) t octads are formed, whereas in the *277 X 138* cross they are not. This means that for t octads to be formed the specific allele (277) would have to be the right (distal) partner in the cross (polarity). For this reason, Whitehouse (1967a) suggested the idea of secondary crossing-over (Fig. 80), which is based on the assumption that the specific alleles (evidently deletions) inducing reciprocal recombination prevent hybridization (see page 169) and are not incorporated into hybrid DNA (they are unable to form molecular heterozygotes). "The details are uncertain, but the apparent

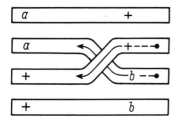

Fig. 80. Diagram of secondary crossing-over within the gene (after Whitehouse, 1967a). Continuous line ending in an arrow denotes hybrid DNA; broken line denotes hybrid DNA eliminated from site *b* before the operator (shown by a dot). All four chromatids are shown.

result is that there is hybrid DNA on the far side of the mutant from the operating point, that is, between the sites of the two allelic mutants, while the parental situation, without hybrid DNA, is restored on the near side, that is, between the site of the first mutant and the opening point" (Whitehouse, 1967a, p. 1355).

To explain the absence of P tetrads it is postulated that the enzyme of reverse crossing-over (see page 176) can initiate the process of dissociation on the other side of the operator only if hybrid DNA is present near it. Pt tetrads can be formed only as a result of reciprocal conversion. Since the frequency of their appearance is negligible (3.8), the strand coefficient in all systems is close to 0.

In Whitehouse's model the presence of hybrid DNA in both chromatids in the immediate neighborhood of the operator is essential. Otherwise crossing-over would be impossible because the conditions for interaction between the chromatids would be absent. Consequently, the strand coefficient ought to be site-specific, like the formation of t tetrads. This is not observed. Mutant *138* is located at the beginning of series *46,* but nevertheless no t octads appear in crosses in which it participates (3.4). The concrete mechanism of formation of hybrid DNA between the sites in secondary crossing-over is uncertain. The hypothesis is based on the assumption that hybrid DNA does not spread as far as the operator although the process of its formation is initiated normally. This presupposes elimination of hybrid DNA from the operator as far as the site of the first mutation. Secondary breakages between the sites are initiated by the ends of fragments of hybrid DNA. A more realistic hypothesis (Paszewski, 1970, personal communication) is that there are only points of predominant primary breakages.

Interpretation of Flank Classes. Whitehouse and Hastings introduced the term "site coefficient" to denote the probability that both sites occur simultaneously in the hybrid region. The site coefficient can be determined by the following reasoning. It is postulated that R classes in a standard flank system (2.8) are formed as a result of the following events (Fig. 81). The R_1 class may appear as the result of complete crossing-over between alleles, in which case the region of hybrid DNA lies between them and does not affect either of the sites (see above). The second possibility is that crossing-over is initiated at the proximal end of the gene and the hybrid DNA occupies only the proximal site a. Correction toward + then gives the R_1 genotype. The third possibility is that crossing-over takes place at the distal end of the polaron and the distal site b, incorporated into the hybrid DNA, is corrected (Fig. 81, A and B). The R_2 class arises because the crossover hybrid DNA occupies both sites which are then corrected toward the wild type (Fig. 81, D).

According to Whitehouse and Hastings (1965) the R_1 class can also arise if the hybrid DNA occupies both mutant sites in the crossover chro-

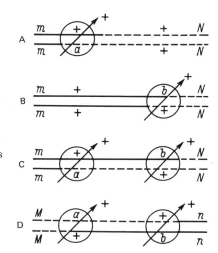

Fig. 81. Mechanism of formation of R classes in a standard flank system in accordance with the polaron-operator model of crossing-over (after Whitehouse and Hastings, 1965). Continuous and broken lines represent DNA strands from different parents. Remainder of legend as in Fig. 72. In each type only the chromatid in which the correction event can lead to the formation of a wild-type recombinant is shown.

matid and they are then both corrected to the wild type (Fig. 81, C). However, if the classes mentioned above in fact arose in this way, the frequencies of appearance of R_1 and R_2 would be the same, which is not observed in reality. It thus follows that the R_1 class arises mainly through the formation of hybrid DNA at one site only.

When crossover hybrid DNA is formed at both sites, R_1 and R_2 classes arise in equal frequencies. As a result of the formation of crossover hybrid DNA at only one site, the R_1 class is obtained. The measure of the frequency of formation of hybrid DNA at both sites is thus $2R_2$ (twice the frequency of class R_2), and the measure of the frequency of formation of hybrid DNA at only one site is $R_1 - R_2$. Consequently, the expression $2R_2/(R_1 - R_2)$ is the site coefficient for a given pair of alleles.

This analysis is based on the erroneous assumption that sites of heterozygosity in hybrid DNA are corrected independently of one another. It will be shown below that *cis* conversion* in *trans*-two-point crosses take place exceptionally rarely, because of the phenomenon of linked correction. The frequency of appearance of the R_2 class cannot therefore be a measure of the formation of hybrid DNA at both sites simultaneously, and the site coefficient proposed by Whitehouse and Hastings can have no definite meaning. It seems that the site coefficient can be determined only by comparing the frequencies of double and single conversions (5.6), but the value of this type of analysis is doubtful in view of the limited data.†

In Whitehouse and Hastings' model double crossovers are still essential to explain the origin of the P configuration of the flanks. The presence of a

---

*Conversions arising in one direction simultaneously $(ab \to +)$ or $(+ \to ab)$.
†The site coefficient can also be determined by comparison of *fk* and *rf* (see page 190).

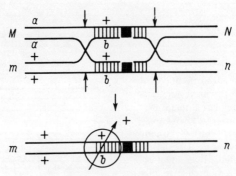

Fig. 82. Mechanism of formation of the P_2 class in a standard flank system (after Whitehouse and Hastings, 1965). Only that chromatid in which the correction event can lead to the formation of a wild-type recombinant is shown below. Arrows indicate sites of secondary breakages. Shaded areas represent hybrid DNA. Remainder of legend as in Fig. 72.

special enzyme, identifying the hybrid configuration of DNA in the operator and initiating the process of dissociation in the neighboring gene, is postulated; this dissociation leads to the formation of a (very short) hybrid DNA in that gene, resulting in reverse (double) crossing-over. The P_1 class arises as the result of correction of the proximal site, while the P_2 class arises through correction of the distal site in the noncrossover (or, more precisely, the double-crossover) hybrid DNA to the wild type (Fig. 82). The results obtained for a system of two adjacent genes (*me-7* and *me-9*) of *Neurospora crassa* (Murray, 1970) do not agree with the mechanism of formation of the P classes described above.

The evident contradictions inherent in the hypotheses described above and their inability to explain adequately many of the experimental facts have compelled a search for other approaches to the solution of the existing (5.1) problems. Let us, therefore, leave aside tetrad analysis for a while and turn to the transformation of bacteria which, in the writer's opinion, can serve as a model for conversion.

5.6. A Theory of Genetic Recombination Based on the Principle of Directed Correction (Kushev, 1970)

The Principle of Directed Correction. The experiments of Ephrussi-Taylor and Gray (2.12) showed that in all mutations (except spontaneous and those induced by proflavine) the integration efficiency (IE) is low, being only about 0.1 of the IE of a standard marker (spontaneous mutation of resistance to streptomycin). Furthermore, IE of the markers was independent of the direction of crossing. It thus followed (Ephrussi-Taylor and Gray, 1966) that LE alleles are corrected to the recipient text from 10 to 20 times more frequently than to the donor, regardless of which allele (+ or −) is transmitted by the fragment of donor DNA. Consequently, the ratio between p^+ and p^- depends mainly, not on the character of the mutation damage, but on which allele in the molecular heterozygote is the recipient. This extremely important property of correction can be called "selectivity" or "directivity." Bresler (1969, personal communication) and Whitehouse (1969) postulated that the selectivity of correction can be determined by breakages in the recombining strand. Another explanation, based on results obtained by Spatz and Trautner (see page 206), is that disturbances of complementation in recombining chromatids are so coordinated that correction takes place most frequently to the recipient text. Vinetskii (1972, personal communication) suggested that the nucleotide sequence established by selection is more resistant from the energetic point of view, and, accordingly, if complementation is disturbed the bases are turned inside out, so that "weak" pairs are formed with their neighbors, and this determines the direction of correction. Correcting enzyme, once it reaches a breakage, is much more likely to utilize the undamaged (recipient) strand as the template for repair. In the case of HE markers selectivity disappears (it may well be that it does not disappear completely, for the standard marker, whose efficiency of integration is taken as 1, can also be corrected). HE markers also are poorly corrected to the donor text, as a result of which they become homozygous only after replication and they form mixed clones (page 214). Many HE markers are true deletions, and acridine mutations often behave as deletions during mapping. Correction of the genetic text of the slave-repeats in accordance with the text of the basic gene (4.2) must obey the principle of directivity (Callan, 1967).

The Model. All the hypotheses of genetic recombination discussed above have been based on the assumption of equality of the p parameters in the two recombining chromatids.* Hence it followed that the strand coefficient must be close to 0. In other words, nonreciprocity of intragenic recombination was regarded as due to the asymmetrical (nonreciprocal) character of formation of hybrid DNA in the recombining chromatids.

With exceptance of the principle of directivity of correction the need for assuming unequal participation of the two chromatids in the formation of hybrid DNA disappears (page 179). It would seem more natural for hybrid DNA to be formed in recombining chromatids by a strictly reciprocal mechanism. Recent findings (Leblon and Rossignol, 1973) confirm this view and cannot be reconciled with any of the types of transformation-like

---

*A step forward was made by Emerson (page 162) but it did not lead to any qualitative progress.

models. The simplest mechanism providing reciprocity is that of half-chiasma formation proposed by Holliday (page 159) and elaborated on the molecular model by Sigal and Alberts (1972).

However, the model now suggested differs from Holliday's model in a number of very significant features.

1. The probability of conversion of the half-chiasma into a whole chiasma is taken to be much less than 0.5 (0.10–0.05).*

2. During the change from half to whole chiasma secondary breakages do not necessarily arise in strictly homologous sites. The participation of repairing enzymes is accordingly necessary for crossing-over to proceed to completion.

3. The recombiners are located outside the gene. One recombiner can serve a group of adjacent genes.

4. In the region of the hybrid DNA (in the half-chiasma) one strand can be regarded as donor (coming from the homologous chromatid), and the other as recipient. Differentiation is only possible, evidently, because of the breakages in the donor strand. Clearly it is impossible to distinguish between the donor and recipient texts in the chiasma (for breakages are present in both strands).

5. Correlation between molecular heterozygosity in the half-chiasma thus obeys the principle of directivity (toward the recipient, i.e., toward the DNA strand unaffected by breakages).

6. By analogy with transformation systems markers can be subdivided into two principal classes: HE and LE (spontaneous and acriflavine markers may have intermediate IE values). Correction for LE markers is strictly directed, while for HE markers directivity has completely disappeared. The degree of directivity of correction for markers with intermediate IE values (ME markers) is reduced to some extent.

The Origin of t Tetrads. Let us assume that during crossing of LE mutants in fungi correction to the recipient text takes place ten times more often than to the donor text, just as during transformation. In that case the table of probabilities of appearance of the various types of tetrads must be drawn up with allowance for the principle of directivity. For simplicity, let us assume that $p^0 = 0$ (as is the case for most mutants).

As Table 15 shows, calculations based on the directivity principle lead to the conclusion that c tetrads are formed ten times more often than t tetrads. This is the same ratio as is observed in crosses of LE mutants (3.4).

The absence of correction leading to postmeiotic segregation means that mutation induces a structural disturbance in the hybrid region which

---

*In the model of Sigal and Alberts there is an elegant possibility of transition from state 3 (Fig. 70) to state 5 with a probability of 0.5. However, since this requires the rotation of one chromosome around the other and, in addition, no cytologically recordable chiasmata are formed as a result, I consider that this mechanism does not operate *in vivo*.

TABLE 15. Probabilities of Appearance of Various Types of Tetrads in Interallelic Crosses Involving LE Mutants, Calculated on the Basis of the Principle of Directivity of Correction

Parameters of correction in chromatid 2	Parameters of correction in chromatid 1	
	$p_1^+ = 0.09$	$p_1^- = 0.91$
$p_2^+ = 0.91$	0.0819 6+ : 2− (c)	0.8281 4+ : 4− (P)
$p_2^- = 0.09$	0.0081 4+ : 4− (t)	0.0819 2+ : 6− (c)

can be identified or corrected only with difficulty. Moreover, the disturbance of the usually rigid mechanism of correction may lead to loss of directivity. In that case $p_1^+ = p_2^+$ and $p_1^- = p_2^-$ and, consequently, the frequency of appearance of t octads may reach 50%. In some cases the frequency of appearance of t octads may exceed 50% (Baranowska, 1970). They may indicate inversion of directivity: the recipient nucleotide sequence is excised more often than the donor.* The reason for the sharp differences in the frequency of appearance of t octads in *cis*- and *trans*-two-point crosses in *Ascobolus* (Paszewski and Prazmo, 1969; Baranowska, 1970), if it cannot be attributed to genetic heterogeneity of the strains, remains an enigma.

It is possible, on the basis of this hypothesis, to explain the absence of 6+ : 2− octads when the h_{3a} mutant is crossed with the wild type (Fields and Olive, 1967). It seems perfectly permissible that in some cases the directivity of correction may be almost absolute. Correction to the recipient text in one chromatid and its random absence in the other gives segregation of 3 : 5 or 5 : 3. Since the parameters p_1^+ and p_2^- have the value 0, no 6+ : 2− and 2+ : 6− tetrads will appear.†

Evidence for the Directed Correction Hypothesis. Quantitative support for the hypothesis of directed correction has been obtained. Gajewski et al. (1968) determined the frequency of appearance of asci of different types in crosses of mutant *277* (from series *46* of *Ascobolus*) with the wild type. These results are given in Table 16.

Determination of the *p* parameters from these results, as Table 16 shows, can be done only by Emerson's method. In this case the value of p^+ is

---
*For another explanation of this fact, see page 194.
†In this case wild-type recombinants can appear in two-point crosses only because of the absence of correction, i.e., in asci of the 1+ : 7− type.

TABLE 16. Comparison of Calculated Expectancies of Different Types of Asci with Experimental Results Obtained in Crosses of Mutant *277* with the Wild Type

Type of ascus	Number found	Number expected	
		a	b
6—:2+	390	370	211
2—:6+	390	370	566
3—:5+	233	244	211
5—:3+	97	99	122

Note: a) Calculated by Emerson's method:

	Chromatid 1	Chromatid 2
p^+	0.178	0.784
p^-	0.712	0.196
p^0	0.110	0.020

b) Calculated by Whitehouse's method ($p_1 = p_2$):

p^+ 0.559
p^- 0.340
p^0 . . . . 0.100

five times smaller in chromatid 1 than in chromatid 2. This is exactly the same ratio as is found in ME mutations (with weakened directivity of correction), so as to give about 20% of t octads in crosses with any other mutations of this series (Table 17).

The theory thus correctly predicts the calculated expectancy of t octads obtained by analysis of the probabilities of appearance of c octads.

The various types of recombinant asci appear in the progeny from crossing another mutant (*1216*) from series *46* with the wild type with prac-

TABLE 17. Analysis of t Octads Obtained by Crossing Mutant *277* from Series *46* with Mutants of This Same Series (Lissouba et al., 1962)

Cross	c asci	t asci
137 × *277*	17	0
63 × *277*	7	4
46 × *277*	6	0
w × *277*	10	2
1216 × *277*	8	0
Total . .	48	6

tically equal frequencies (Gajewski et al., 1968):

Type of ascus	Frequency (10^3)
6— : 2+	3.7
6+ : 2—	4.2
5— : 3+	5.0
5+ : 3—	4.4

It will be clear that in this case the degree of directivity of correction cannot be determined. In two-point crosses involving this mutant no t octads can be found, further support for the relativity of correlation between postmeiotic segregation and reciprocal recombination in the gene.

Directivity of correction can also be found by analysis of a flank system. Results of this type have been obtained for the g gene of *Sordaria fimicola* (Kitani and Olive, 1967, 1969). Let us examine the simplest case: a standard flank system with one site of heterozygosity. As Fig. 83 shows, a chromatid in which the correction to + or to — has taken place can be identified in asci with the 5 : 3 segregation. Depending on the configurations of the flanks two main classes can be distinguished (P and R). Correction to the recipient text (+ or) gives subclasses R_r and P_r, correction to the donor text gives subclasses R_d and P_d. Comparison of the results obtained for six allelic h mutations within the classes leads to the conclusion that correction to the recipient text in the P class takes place on the average three times more often than to the donor text (102 cases against 37).* Within the R class, however, this asymmetry is much weaker (88 against 65). This fact is evidence that during formation of a crossover flank configuration the ability to discriminate between the donor and recipient texts is lost, as the proposed hypothesis demands.

The g_1 mutation reduces the directivity of correction to zero (this is the HE allele). In two-point (g_1 × h) crosses wild-type recombinants appear predominantly through conversion of the g_1 allele. Consequently, in these crosses t octads must appear in high frequency (according to the model proposed). This is in fact observed experimentally (Kitani and Olive, 1969): 26 t octads from 46 containing wild-type spores. Moreover, just as in crosses involving g_1, postmeiotic segregation is very frequently observed, and many octads (17 of 26) contain only one double mutant spore (the reciprocal product during formation of the wild type). Clearly the appearance of these octads requires the presence of hybrid DNA in the g_1 site. Consequently, they cannot be caused by secondary crossing-over.

Distribution of Flank Markers. An explanation of why t tetrads usually arise in the R_1 class is called for (Whitehouse, 1967a; the summary table

---

*Kitani and Olive used this fact to explain polarity in terms of a model structurally analogous to that proposed by Taylor (5.3). It will be shown below (page 183) that polarity is the result of linked correction. In an analogous system (Stadler and Towe, 1971) the direction factor of correction reaches 50.

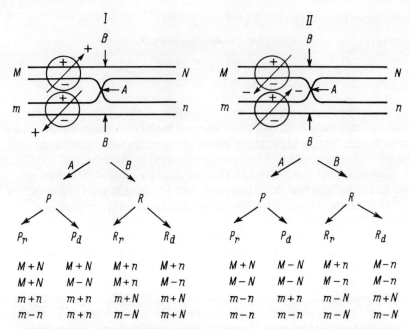

Fig. 83. Determination of the direction of correction in a flank system with one site of heterozygosity in asci of 5:3 type in *Sordaria fimicola* (after Kitani and Olive, 1967): I) correction to the wild type in one chromatid while the other remains heterozygous; II) correction to the mutant type in one chromatid while the other remains heterozygous. A) Breakage of the half-chiasma leading to the formation of P configurations of the flanks; B) breakages leading to R configurations. R_r and P_r classes are formed by correction to the recipient text, R_d and P_d classes by correction to the donor text. Types of asci obtained in crosses between the wild type strain with six alleles and *h* mutants. Only genotypes of spores formed by recombining chromatids are shown.

gives 147 R_1 t tetrads but only 16 t tetrads of all other types). In terms of the proposed model this is easily explained on the assumption that in the chiasma (R_1 class) correction is undirected, and reciprocal conversions take place there with high frequency. It is therefore also postulated that the probability of conversion of the half-chiasma into a whole chiasma is below 0.5 (of the order of 0.1). Otherwise the P configurations would be considerably fewer than the R_1. This argument leads to a highly specific prediction. Mutations of LE type must give t tetrads predominantly in the R_1 class, whereas in crosses involving HE mutations Pt and R_1t tetrads must appear with equal frequencies. In experiments with g_1 (Kitani and Olive, 1969) seven R_1t and two Pt octads were found. The sample was evidently too small, but the tendency for the relative proportion of Pt octads to increase is evident. Detailed analysis of the flank system with different types of mutants provides an excellent test for this model.

In experiments with HE mutations the probability of transition from half-chiasma to whole chiasma can be determined by the ratio

$$\frac{R_1}{P_1 + R_1}$$

For crosses involving the g_1 mutant (of HE type) the value of this ratio is 0.1 (19 R_1 and 184 P_1 octads: Kitani and Olive, 1969, Table 3a, asci with normal distribution for h alleles), as the proposed model demands.

The hypothesis of directed correction is better able to explain variations in the distribution of flank markers than any other hypothesis. Besides differences in the frequencies of dissociations at different ends of the gene which are allowed for in Holliday's and Whitehouse's models, the character of correction of the allele concerned has a very marked effect on the frequency of appearance of the various flank classes (as we have just seen). It must also be remembered that the character of correction of sites of heterozygosity in recombining chromatids is dissimilar (5.4) and, consequently, mutations weakening the directivity of correction can be manifested in only one chromatid. This is shown by analysis of the g_1 mutant. In one chromatid correction is undirected, while in the other it remains directed (5.4; Kitani and Olive, 1969). These differences can also be exhibited between the chiasma and half-chiasma, for a change in the position of the sites of heterozygosity relative to the flank markers during a change of polarity (Fig. 84) is valid only for P configurations.

The fact that semicrossing-over does not interfere with crossing-over (3.8) may indicate that the appearance of chromosome interference is connected with regulation of the action of the endonuclease inducing the secondary breakages.

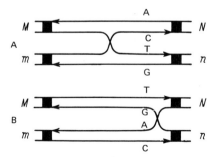

Fig. 84. Change in position of sites of heterozygosity relative to flank markers (after Emerson, 1966b). A) Breakages in distal, and B) in proximal recombiners. Site of heterozygosity A/C lies either in chromatids $MN(A)$ or in chromatid $mn(B)$. Note that if breakages take place during interchange of recombiners in DNA strands of different polarity, no displacement of the sites takes place.

TABLE 18. Possible Alternative Ways of Formation of Aberrant Asci in an $ab \times +$ Cross

	Variant number and events						
	1	2	3	4	5	6	7
	$c\,(+\to a)$	$c\,(+\to b)$	$c\,(+\to a)(+\to b)$	$c\,(a\to +)(b\to +)$	$c\,(a\to +)(+\to b)$	$c\,(+\to a)(b\to +)$	t
Genetic constitution of ascus	ab	ab	ab	ab	ab	ab	ab
	ab	ab	ab	$++$	$+b$	$a+$	$a+$
	$a+$	$+b$	ab	$++$	$+b$	$a+$	$+b$
	$++$	$++$	$++$	$++$	$++$	$++$	$++$
Segregation by genotype	$2\,ab:1\,a+$ $:1++$	$2\,ab:1+b:$ $:1++$	$3\,ab:1++$	$1\,ab:3++$	$1\,ab:2+b:$ $:1++$	$1\,ab:2\,a+:$ $:1++$	$1\,ab:1\,a+:$ $:1+b:$ $:1++$
Segregation by phenotype	$1+:3-$	$1+:3-$	$1+:3-$	$3+:1-$	$1+:3-$	$1+:3-$	$1+:3-$

Polarity. In correction models of recombination polarity is explained by assuming that the formation of hybrid DNA (the dissociation double-association cycle) begins predominantly at one of the gene. Consequently, it is more probable that the site located nearer to the recombiner lies in the hybrid region that its partner (in a two-point cross). However, polarity cannot be explained on the basis of this principle alone. In fact, the question may be asked: why do wild-type recombinants appear in a two-point cross entirely by conversion of only one of the alleles and not of both at the same time? The probability of simultaneous incorporation of closely linked markers into hybrid DNA ought to be higher than the probability of incorporation of one of them. Clearly the answer to this question would at the same time provide the key also to the understanding of the linear relationships within the gene (i.e, the linear dependence of rf on physical distance).

To analyze this problem let us consider the possible alternative ways in which aberrant asci are formed when a double mutant (ab) is crossed with the wild type (Table 18). In the progeny from such a cross the appearance of asci with segregations of $1+ : 3-$ and $3+ : 1-$ will be expected. The frequency of formation of asci of the last type (variant 4) must be given by the product $fk_{(a \to +)} \times fk_{(b \to +)}$ if the conversion events at the sites a and b are independent.

The frequency of appearance of asci with the segregation $1+ : 3-$ is determined in a more complex fashion: they can be formed either as a result of single conversions (variants 1 and 2), of double conversions of various

TABLE 19. Frequency of Formation of Asci with the Segregation $3+ : 1-$ in $ab \times +$ Crosses Involving Mutants from Series *19* (Mousseau, 1966) and Series *y* of *Ascobolus* (Kruszewska and Gajewski, 1967)

Series	Cross $a \quad b \times +$		$fk_{(x \to +)} \cdot 10^3$		$fk_{(a \to +)(b \to +)} \cdot 10^3$	
			$x = a$	$x = b$	Theoretically expected $fk_{(a \to +)} \times fk_{(b \to +)}$	Experimental
19	A270	C55	2.4	2.0	0.0009	0.24
	A60	C55	1.0	2.0	0.0018	0.64
	A1028	C55	0.5	2.0	0.0004	0.33
	A1028	B1844	0.5	5.0	0.0034	0.41
	A2073	B19	1.2	0.6	0.0009	0.59
y	y	146	0.03	1.55	0.000047	0.03 ± 0.03
	y	77	0.03	2.50	0.000075	0.11
	y	775	0.03	1.09	0.000033	0.06 ± 0.04
	794	77	1.23	2.50	0.003140	0.27 ± 0.10
	794	775	1.23	1.09	0.001340	0.30 ± 0.12

TABLE 20. Frequency of Formation of Asci with the Segregation
$3-:1+$ in $ab \times +$ Type Involving Mutants from Series *19* of
Ascobolus (after Mousseau, 1966)

Cross $ab \times +$		$fk_{(+ \to x)} \cdot 10^3$		Frequency of appearance of octads $(3-:1+) \cdot 10^3$	
		$x = a$	$x = b$	Theoretically expected	Experimental
A270	*C55*	5.0	4.4	9.4	0.20
A60	*C55*	0.9	4.4	5.3	0.57
A1028	*C55*	0.2	4.4	4.6	0.24
A1028	*B1844*	0.2	3.5	3.7	0.45
A2073	*B19*	0.4	0.4	0.8	0.31

types (3, 5, 6), and of reciprocal recombination (7). The appearance of the different types of asci is recorded visually, and in experiments of this sort it is therefore easy to collect a large enough material. The results obtained (Mousseau, 1966; Kruszewska and Gajewski, 1967) are given in Table 19. The frequencies of simultaneous conversions at two sites, as this table shows, are hundreds or thousand times greater than those expected on the assumption that these events are independent. Comparison of $fk_{(x \to +)}$ with $fk_{(a \to +)(b \to +)}$ shows that the values of this last term are determined by $fk_{(x \to +)}$ of that site of the pair for which this value is the smaller, viz., $fk_{(x \to +)\text{min}}$.

For all crosses except $A270\ C55 \times +$ in series *19* and $79477 \times +$ and $794775 \times +$ in series *y*, in which it is 5-10 times smaller, $fk_{(a \to +)(b \to +)} \simeq fk_{(x \to +)\text{min}}$. It is interesting to note that a similar situation is observed during crossing of extreme sites, for which $fk_{(x \to +)}$ values are very high and approximately equal. There is thus statistically significant evidence that double conversions are interlinked events.

As Table 18 shows, the frequency of appearance of asci with the segregation $3-:1+$ in a *cis*-two-point cross is determined by the frequencies of appearance of variants 1, 2, 3, 5, 6, and 7. This means that the frequency of appearance of asci of this type must be determined at least by the sum of $fk_{(+ \to a)}$ and $fk_{(+ \to b)}$. Is this prediction valid?

The results given in Table 20 show that it is not. The frequency of formation of $3-:1+$ octads is not determined by the sum of $fk_{(+ \to a)}$ and $fk_{(+ \to b)}$, but like the frequency of appearance of $3+:1-$ octads (Table 19), it is determined by $fk_{(+ \to x)\text{min}}$. Hence it follows that, in the presence of two alleles, conversion takes place simultaneously in both sites and that the frequency of simultaneous conversion is determined by the allele with the smaller fk value. To verify these conclusions let us examine the results obtained by genetic analysis of 3− : 1+ asci. By crossing the double mutant *1604 137* with the wild type (Rizet and Rossignol, 1963) 32 octads with the segregation 6− : 2+ were obtained, and among them the following types were found:

12 octads 3 (*1604 137*) : 1 (+ +) of variant 3,
19 octads 2 (*1604 137*) : 1 (*137* +) : 1 (+ +), variant 2,
1 octad 1 (*1604 137*) : 1 (*137* : 1 (*1604*) : 1 (+ +), variant 7
(Table 18).

The frequency of simultaneous conversions (variant 3) is comparable in value with $fk_{(+ \to 1604)}$, which is approximately half that of $fk_{(+ \to 137)}$. Single conversions (+ → *1604*) are not found, but nevertheless more than half of all the asci were formed as a result of the conversion (+ → *137*). The absence of variants 5 and 6 will be noted again. The frequency of appearance of t octads is negligible. Hence, a feature of *cis*-two-point crosses is the absence (or very low frequency) of formation of variants 5 and 6, i.e., of asci whose formation requires simultaneous conversions in different directions (*trans* conversion); simultaneous conversions in the same direction (*cis* conversion) are found at frequencies determined by fk_{min}. Since the basic frequency of conversion of an allele reflects the probability of hybrid DNA formation in that particular site it must be concluded that if hybrid DNA, as it spreads from the recombiner, occupies both mutant sites, *cis* conversion (correction in both sites in the same direction simultaneously) takes place. Proof of the existence of linked correction has also been obtained in experiments of another type.

Let us examine the two-point cross $a+ \times +b$ (Table 21), in which recombination takes place entirely as a result of conversion (of one site or of both simultaneously). Phenotypically nonrecombinant asci will also be considered if conversion occurred in them. Variants 1, 2, and 9 will be recombinant tetrads whose total frequency determines *rf* between the alleles under selective conditions. The relationship between *rf* and *fk* will be examined below (page 190). Judging from the fact that the frequency of appearance of asci of 4 : 4 type (Table 21, variant 3) is extremely low (Lissouba et al., 1962), it may be supposed that the frequency of simultaneous conversions of two alleles is given by the product of the probabilities. Let us consider, however, what is happening with phenotypically normal asci (Table 21, variants 4-8). An analysis of octads of this type was undertaken (Rizet and Rossignol, 1963) in order to study double mutants. Crosses *62* × *137* (series *56*) were carried out and asci containing only white mutant spores were analyzed. Altogether 750 asci were analyzed, and an aberrant segregation was found in 9 of them. The genotypes of the spores in these octads are given in Table 22.

TABLE 21. Possible Alternative Ways of Formation of Aberrant and Phenotypically Normal Asci in $a+ \times +b$ Crosses

	Variant number and events								
	1	2	3	4	5	6	7	8	9
	$c(a \to +)$	$c(b \to +)$	$c(a \to +) \times$ $c(b \to +)$	$c(+ \to a)$	$c(+ \to b)$	$c(+ \to a) \times$ $c(+ \to b)$	$c(a \to +) \times$ $c(+ \to b)$	$c(+ \to a) \times$ $c(b \to +)$	t
Genetic constitution of ascus	$a+$ $++$ $+b$ $+b$	$a+$ $a+$ $++$ $+b$	$a+$ $++$ $++$ $+b$	$a+$ $a+$ ab $+b$	$a+$ ab $+b$ $+b$	$a+$ ab ab $+b$	$a+$ $+b$ $+b$ $+b$	$a+$ $a+$ $a+$ $+b$	$a+$ $++$ ab $+b$
Segregation by genotype	$1a+:$ $2+b:$ $1++$	$2a+:$ $1++:$ $1+b$	$1a+:$ $2++:$ $1b+$	$2a+:$ $1ab:$ $1+b$	$1a+:$ $1ab:$ $2+b$	$1a+:$ $2ab:$ $1+b$	$1a+:$ $3+b$	$3a+:$ $1+b$	$1a+:$ $1++:$ $1ab:$ $1+b$
Segregation by phenotype	$1+:3-$	$1+:3-$	$2+:2-$	$4-:0+$	$4-:0+$	$4-:0+$	$4-:0+$	$4-:0+$	$1+:3-$

TABLE 22. Types of Octads Containing Only White Spores Obtained in a *63 × 137* Cross (after Rizet and Rossignol, 1963)

Genotypes of spore pairs in asci	Number of octads	Variant (Table 21)
2 (*63*) : 2 (*137*)	741	—
3 (*63*) : 1 (*137*)	5	8
1 (*63*) : 3 (*137*)	1	7
1 (*63*) : 2 (*137*) : 1 (*63 137*)	1	5
2 (*63*) : 1 (*137*) : 1 (*63 137*)	2	4

Unfortunately because of technical difficulties it was impossible to obtain an extensive statistical material, but this example alone will suffice to show that simultaneous conversions take place even more often than single conversions. On the other hand, absence of variant 6 is striking.

Fogel and Mortimer (1969) recently published the results of an analysis of more than 2000 unselected tetrads obtained from crosses involving four alleles of the arg_4 gene and two alleles of the leu_1 gene of *Saccharomyces cerevisiae* (Table 23). It is clear from Table 23 that *trans* conversion types of asci are readily found in *trans*-two-point crosses, whereas *cis* conversions (Table 21, variants 3 and 4) are completely absent (or occur at lower frequency: for example, in a study of recombinations within the *6A* region of *Sordaria brevicollis,* of a total number of 255 c tetrads only 4 were found to contain *trans* conversions; Fields and Olive, 1967).

These facts are evidently well described in the terms of linked correction (see page 214). If both markers are in the hybrid region in the *trans* position, correction of one site of heterozygosity to + leads to correction of

TABLE 23. Analysis of Unselected Tetrads Obtained from Crosses Involving Various Alleles of Genes arg_4 and leu_1 of *Saccharomyces cerevisiae* (after Fogel and Mortimer, 1969)

Events in the gene	Distance between alleles, in nucleotides			
	arg_4 4×17 1080	arg_4 1×2 520	arg_4 2×17 128	$leu_1 \times leu_{12}$ 1030
c $(a \to +)$	3	3	1	4
c $(+ \to a)$	5	3	3	6
c $(b \to +)$	18	10	3	0
c $(+ \to b)$	20	11	2	0
c $(+ \to a)(b \to +)$	1	10	13	1
c $(a \to +)(+ \to b)$	2	13	14	1
t	9	5	0	0
Σ	58	55	36	12

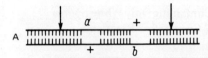
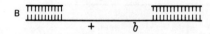
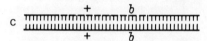

Fig. 85. Scheme of simultaneous correction of two sites of heterozygosity in the DNA molecule. Arrows mark boundaries of excised region; broken line indicates resynthesized material. A-C) Stages of correction.

the other site of heterozygosity to − (Fig. 85). In this way simultaneous correction of both sites to + (the two-site "associated conversion" of Stadler and Kariya, 1969; Paszewski et al., 1971) is a rare event and is determined by the probability of absence of linked correction (see page 192). The coefficient $\Delta = fk_{ab} \to +/fk_{(x \to +)\min}$ defines the probability of linked correction.

The above analysis shows that in *trans*-two-point crosses prototrophs can be formed, because of linked correction, only through conversion of one of the sites (distal or proximal). Consequently, hybrid DNA must end between these sites. Polarity thus arises in two-point crosses because of selection for this situation.

Linear Measurements. Correction models of recombination met with a very serious difficulty for they were unable to explain why recombination frequencies in two-point crosses can be used for drawing linear maps of genes. The *rf* value of a pair of mutations in a gene depends on the frequency with which any site is incorporated into hybrid DNA. The farther apart the mutations, the closer they are to the ends of the gene and, consequently, the more often any one of them (if the formation of hybrid DNA can start at either end of the gene) is incorporated into the hybrid region and the higher the value of *rf* between them. The *rf* value thus to some extent reflects the actual physical distance between the alleles and it can be used for mapping the gene. The difficulty arises in the case of crosses of closely linked mutants: the closer they are to each other the greater the chance that they will be simultanesouly incoporated into the hybrid region. If the two sites of heterozygosity are corrected to the wild type independently, these events are independent of the distance between the sites. For this reason, crosses between closely linked mutants will give *rf* values comparable with those between distant mutants. Holliday attempted to get around this difficulty by suggesting that mutant sites themselves prevent the pairing of complementary strands. Association of strands requires strict homology; if the homology is diminished the effectiveness of pairing is also reduced. The closer the sites, the more strongly they interfere with pairing. In this case a linear map of the gene can be drawn. However, strict additivity can be obtained only if all mutants inhibit pairing equally, which is improbable. Moreover, in short segments the intensity of inhibition may increase, not as a linear function of distance, but more sharply. This may have the effect of widening the map. However, it is difficult to understand how the effect of inhibition can spread over distances of hundreds of nucleotides.

All these difficulties can be overcome by the principle of linked correction. As I showed in the last section, wild-type recombinants are formed in *trans*-two-point crosses only if the hybrid DNA ends between the sites. It is evident that the probability of formation of the original half-chiasma is a function of distance. This idea was first expressed by Pritchard in 1964 (cited by Holliday, 1966a).

Hence, if the frequency of t tetrads in intragenic ($a \times b$) crosses can be disregarded, rf_{ab} must be determined by the equation (Kruszewska and Gajewski, 1967):

$$4rf_{ab} = fk_{(a \to +)} + fk_{(b \to +)} - x, \qquad (1)$$

where

$$x = fk_{(a \to +)(+ \to b)} + fk_{(b \to +)(+ \to a)}.$$

When *rf* is determined by means of this equation the following experimental facts must be taken into account:

1. In $a \times b$ crosses conversion affects predominantly the allele at the beginning of the polaron.
2. The frequencies of linked *trans*-conversions $fk_{(a \to +)(+ \to b)}$ and $fk_{(+ \to a)(b \to +)}$ are determined by fk_{min}. In the absence of allele specificity fk_{min} is characteristic of sites at the end of the polaron or, if the gene is bipolar, in the middle of the gene.
3. In all genes grouping of the sites is found. In series *y* there are two groups: alleles *794, y,* and *183* at the left end of the series, and *146* and the other alleles at the right end. In series *46* there are also two groups: alleles *137* and *237* at the right end (the beginning of the polaron) and the remainder in the left. Series *19* consists of two groups (A-B and C).
4. Groups of sites may have the same (series *46*) or different (series *19* and *y*) polarity.

For a unipolar gene (for example, gene *46* in *Ascobolus*) with distal polarity, $fk_{(b \to +)} > fk_{(a \to +)}$, and consequently

$$fk_{(a \to +)(+ \to b)} = fk_{(a \to +)}.$$

TABLE 24. Determination of rf between Mutants of Series 46 of *Ascobolus* on the Basis of Equation (2) (after Lissouba et al., 1962)

Cross $a \times b$	$fk_{(b \to +)} \cdot 10^3$	$fk_{(+ \to a)} \cdot 10^3$	$4\, rf_{ab} \cdot 10^3$	
			Theoretically expected	Observed
46×63	4.54±0.35	1.90±0.60	1.69—3.59	0.53
$w \times 63$	4.54±0.35	3.87±0.76	0.00—1.78	2.82
$w \times 46$	4.94±0.95	3.87±0.76	0.88—2.89	0.74

Equation (1) acquires the form

$$4\, rf_{ab} = fk_{(b \to +)} - fk_{(+ \to a)(b \to +)}. \tag{2}$$

This equation is in good agreement with the experimental data (Table 24), especially considering that the determination of $fk_{(+ \to x)}$ for *Ascobolus* is beset by technical difficulties (Kruszewska and Gajewski, 1967) and the values obtained are not always very accurate.

Expansion of the Map. A possible explanation of the effect of expansion of the map was suggested by Pritchard (cited by Holliday, 1968). Let us assume that during correction of a site of heterozygosity not less than n nucleotides are as a rule removed. If the distance between the markers exceeds n nucleotides wild-type recombinants may also appear when both sites lie in the region of hybrid DNA. In this case independent correction (i.e., *cis* conversion) is possible because linked correction will not take place in 100% of cases. It thus follows that expansion of the map is observed at distances greater than n. Within a region of the map measuring n nucleotides no expansion will occur because the contribution of the second component to rf is negligible. If we add together the rf values obtained for segments smaller than n, we have to deal only with the first component. It is evident that rf between markers several multiples of n apart will be much greater than the sum of the intermediate rf values (Fincham and Holliday, 1970).

The results obtained by a study of recombination in the ad_6 gene of *Schizosaccharomyces* (Leopold and Gutz, cited by Holliday, 1968) give a value of $n = 65\text{-}130$ nucleotides. This figure (hundreds of nucleotides) is an order of magnitude below that obtained in transformation systems. The calculation is based on the assumption that exonucleotic degradation can take place from both the 3'- and 5'-end of the gap. Analysis of conversions in *cis*-two-point crosses shows that linked correction may extend to the entire gene, and even to neighboring genes (Paszewski, 1967; Fogel et al., 1970). The length of the DNA region excised during correction may evidently amount to several thousands of nucleotides.

In bipolar genes the expansion effect may also take place because the process of dissociation, which is initiated at one end, very rarely reaches the other end. Accordingly, if the crossed mutants belong to different groups with opposite polarities the value of x may be disregarded (see Table 19, series y). Equation (1) then acquires the form

$$4\,rf_{ab} = fk_{(a \to +)} + fk_{(b \to +)}.$$

Determination of rf by this equation gives very good agreement between the theoretically expected and observed values of rf (Table 25).

This evidently also explains the site grouping effect in bipolar genes (series *19* of *Ascobolus* and the *cys* gene of *Neurospora crassa*; Stadler and Towe, 1963).

When the expansion effect is slight or absent altogether the situation is evidently one of uncorrected mutations, a small gene, or an ineffective correction system.

Allele Specificity of Recombination. One very important result follows from the proposed model of genetic recombination: the value of rf determined in two-point crosses is dependent on the degree of directivity of correction. As Fig. 73C shows, a "prototrophic" chromatid arises through correction of a site of heterozygosity in the *mn* chromatid to the wild type. Since the wild-type allele is the donor, $p^+_{mn} \ll p^-_{mn}$, if the *b* marker has low integration efficiency (LE). The *b'* homoallele, with high integration efficiency (HE) gives $p^+_{mn} = p^-_{mn}$. Consequently, $rf_{ab'}$ is much higher than rf_{ab}. A change in the scale of the map of the *ami A* locus in *Pneumococcus* if different groups of mutants (HE or LE) are used is a clear example of relationships of this type between the recombination frequency and the character of mutation injury. The contribution of crossing-over and semicrossing-over to rf between markers is thus allele-specific: this contribution for LE

TABLE 25. Determination of rf_{ab} between Series y Mutants of *Ascobolus* on the Basis of Equation (3) (from the data of Kruszewska and Gajewski, 1967)

Cross $a \times b$	$fk_{(a \to +)} \cdot 10^3$	$fk_{(b \to +)} \cdot 10^3$	$4\,rf_{a\,b} \cdot 10^3$	
			Theoretically expected	Observed
794 × 146	1.23	1.55	2.78	2.64
794 × 73	1.23	0.09	1.32	1.39
794 × 77	1.23	2.50	2.73	3.09
794 × 775	1.23	1.09	2.32	2.27
183 × 146	0.06	1.55	1.61	0.66
183 × 73	0.06	0.09	0.15	0.06
183 × 77	0.06	2.50	2.56	2.37
183 × 775	0.06	1.09	1.15	0.79
y × 146	0.03	1.55	1.58	1.67
y × 73	0.03	0.09	0.12	0.30
y × 77	0.03	2.50	2.53	2.71
y × 775	0.03	1.09	1.12	1.09

mutations is about equal, while in the case of HE mutations wild-type recombinants appear predominantly through semicrossing-over.

Clearly comparable rf values can be obtained only in crosses between mutants with equal IE. Spontaneous mutations which, as the experiments of Gray and Ephrussi-Taylor showed, have a wide IE spectrum are very difficult to map. Mutagens (except x-rays and acridine dyes) induce mutations less heterogeneous with respect to IE. However, even *amber* mutations are not a homogeneous group (Norkin, 1970). Selection of individual groups of homogeneous mutants may have the result that in one case LE mutations are mainly selected for mapping, and HE mutations in another. As a result, the dimensions of the genes in the same organism will differ by an order of magnitude. This is a characteristic situation, evidently, for series *75* and *46* of *Ascobolus immersus*. Series *75* occupies a region of almost 3 map units, whereas gene *46* occupies only 0.3 of a map unit.

The proposed hypothesis also explains the observed correlation between the frequency of appearance of t tetrads and their distance apart on the map. Since rf in interallelic crosses is not always a measure of the physical distance between the sites, a causal relationship must exist between the frequency of appearance of t tetrads and rf, but not vice versa. Detailed analysis of the system is possible if it satisfies certain demands:

1. The order of the markers must not only be determined in two-point crosses, but also verified in three-point crosses. The ideal system, of course, is one with flank markers.
2. The basic frequencies of conversions of all alleles must be strictly determined in one-point crosses.
3. An extensive statistical material covering relationships between the types of tetrads in each cross is essential.
4. The mutants used must be induced by particular mutagens, i.e., they must form homogeneous groups.

Mutants belonging to series *75* of *Ascobolus immersus* are mapped in a segment 3 map units long.* As many as 98% of t octads appear in crosses within the series (Fig. 86). However, it is difficult to interpret the results relating to this series purely on the basis of hypothetical correlation between rf and the frequency of the t octads because of the sharp increase in frequency of the t octads (up to 98%) in crosses between groups of proximal and distal alleles. Crosses within groups (*1186–278* and *2029–1472*) give fewer t octads (not more than 50%). There is yet another difficulty of interpretation. It could be supposed that mutant *1472* belongs to the HE type, which is why it gives so many t octads as well as higher rf values. However,

---

*In series *75* only 12 of the 140 spontaneous mutants have been investigated (Lissouba et al., 1962; Rizet and Rossignol, 1966). During the mapping many difficulties were encountered (allele-specific recombination of homoalleles, nonadditivity of rf values, post meiotic segregation, and so on), and only those alleles which could be arranged in linear order were therefore selected.

§ 5.6] THEORY OF DIRECTED CORRECTION 195

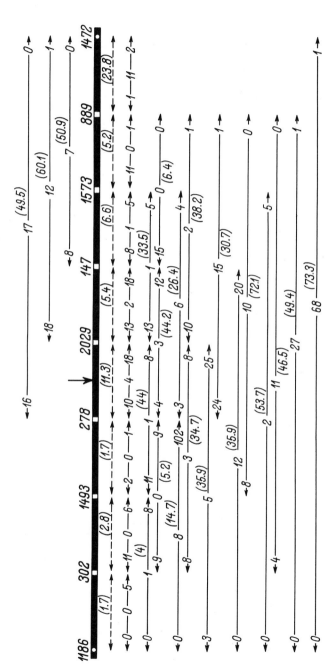

Fig. 86. Map of series 75 of *Ascobolus immersus* (from Rizet and Rossignol, 1966). Thick line represents map of the locus. Empty circles and squares mark positions of mutant alleles. Numbers in parentheses above and below lines terminating in arrows and connecting pairs of mutation sites denote frequency of appearance of recombinant octads (1 × 10³). Numbers inside lines denote number of recombinant octads analyzed in a particular cross (numbers on the left-hand and right-hand sides and in the center of the lines denote numbers of octads with conversion of the proximal and distal site and number of t octads respectively). The bold arrow marks the position of the recombiner.

this conflicts with the polarity observed in c octads. It is only in exceptional cases that c octads are formed as a result of conversion of the *1472* allele. Consequently, polarity is proximal (the recombiner is at the left, proximal end). At the same time, the influence of allele *1472* on recombination could be manifested only if distal polarity were present. These contradictions suggest that series *75* is bipolar, and that the recombiner lies in the middle of the locus (between sites *278* and *2029*).

A way out of the situation thus arising is indicated by determination of the frequencies of conversions in single-site crosses. Rossignol (1969) divided mutations in this gene into four groups (α, β, γ, and δ) on the basis of the coefficient of asymmetry (DC = $fk_+ \to _a/fk_a \to _+$). The corresponding values are: $DC_\alpha = 14$, $DC_\beta = 1$, $DC_\gamma = 0.7$, and $DC_\delta = 0.02$. The polarity of fk (distal) was found within each individual group. Since α mutations are characterized by the lowest values of fk this group of mutants is presumably of the LE type. The β and γ groups include HE and ME mutations. The direction factor in this system is low (3-5). So far as the absence of polarity in two-site crosses is concerned, according to Rossignol "this may reflect the occurrence of frequent independent correction events when the two mutant sites are involved in hybrid DNA" (page 803).

In series *84W* of *Ascobolus immersus* only four alleles have so far been mapped (Paszewski and Prazmo, 1969). The order of the sites (Fig. 87A) was

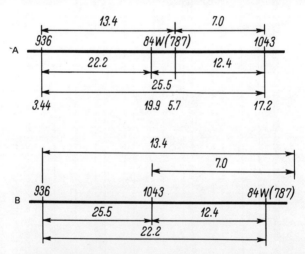

Fig. 87. Map of series *84W* of *Ascobolus immersus*: A) order of the markers determined by three-point crosses (after Paszewski and Prazmo, 1969); B) when the order of the markers was determined the character of polarity was taken into consideration (original). Distances are marked in frequencies of appearance of asci of the 6– : 2+ type (1×10^3) in two-point crosses. Number of homoallele shown in parentheses. Below the vertical lines $fk_x \times 10^3$.

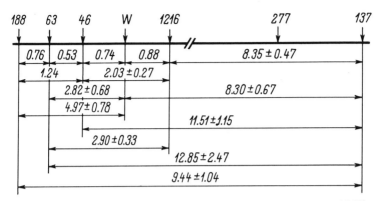

Fig. 88. Map of series *46* of *Ascobolus immersus* (after Lissouba *et al.*, 1962). Distances expressed in frequencies of appearance of (6– : 2+) octads (1 × 10³) in two-point crosses.

confirmed in three-point crosses, but analysis of the polarity shows that alleles *1043* and *84* (*787*) must change places (Fig. 87B).

The results of four-site crosses (Paszewski et al., 1971) agree only with the original picture. As the facts described above demonstrate, polarity in two-site crosses also is evidently not an absolutely reliable criterion for localization of mutations. It must therefore be noted that the *M26* (Gutz, 1971b,c) and *H5*(Kitani, 1972) alleles, even though they distort the polarity, need not necessarily be new recombiners, for the δ allele of Rossignol behaved in exactly the same way.

In series *46* there is only one allele (*277*) with partially weakened directivity of correction (which was shown earlier). This means that it can be localized only in crosses with mutant *137*, which is located more distally (Fig. 88).

Series *y* is interesting because the complementation test was negative when used on all the mutants studied. The map of this gene is shown in Fig. 89. The order of the sites was determined by three-point as well as two-point crosses, nevertheless analysis shows that this order is not satisfactory. The results given in Table 26 show that the *y* gene is bipolar; the homoalleles *y* and *183* lie on the boundary between the two polarons. Meanwhile, at the distal end of the gene polarity is inverted in 77 × 775 crosses. It is therefore more logical to change the places of these alleles. The results of three-point crosses do not conflict with such an arrangement of the sites. As Table 26 shows all mutants in this series except *y* are of the LE type. In 73 × *y* crosses t octads may appear as a result of reciprocal conversions of either allele. However, for several reasons* it is unlikely that allele *73* is of the HE

---

*Allele *73* gives low *rf* values, on account of the low value of $fk_{(73 \rightarrow +)}$(see the map). If, however, allele *73* were of the HE type, the total basic frequency of its conversion would have to be one order of magnitude higher than is actually observed (7 × 10³) instead of 0.7 × 10³.

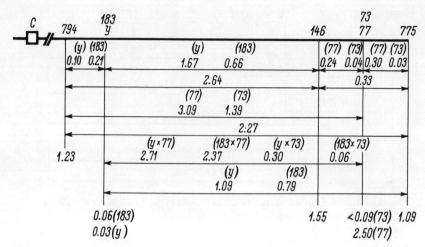

Fig. 89. Map of series y of *Ascobolus immersus* (from Kruszewska and Gajewski, 1967). The order of the mutations was established by three-point crosses. Numbers above the map and in parentheses denote numbers of alleles. Number above the lines terminating in arrows correspond to frequencies of $(6-:2+)$ octads (1×10^3) in corresponding crosses. Numbers below vertical lines represent $fk_{(x \to +)} \times 10^3$ for the corresponding mutants. C = centromere.

TABLE 26. Types of Recombinant Octads Formed by Crosses between Alleles of the y Gene of *Ascobolus* (Kruszewska and Gajewski, 1967)

Cross $a \times b$	Number of octads studied	Octads	c octads		
			total	Conversion in a site $c(a \to +)$	Conversion in b site $c(b \to +)$
$794 \times y$	3	0	3	3	0
794×183	15	0	15	15	0
794×146	39	2	37	8	29
794×77	26	1	25	2	23
794×73	21	2	19	19	0
794×775	30	2	28	12	16
$y \times 146$	27	1	26	0	26
$y \times 77$	30	1	29	0	29
$y \times 73$	9	6	3	1	2
$y \times 775$	31	1	30	0	30
183×146	19	0	19	0	19
183×77	29	0	29	0	29
183×775	31	2	29	0	29
146×77	9	0	9	1	8
146×775	14	0	14	5	9
77×775	11	0	11	9	2
Total . . .	344	18	326	75	251

type. The reason for the low value of fk_y is that the y site lies on the boundary between two polarons and hybrid DNA reaches it only one-seventh as frequently as it reaches site 73 (77).

The Genetic Control of Intragenic Recombination. The hypotheses of Holliday and of Whitehouse and Hastings cannot explain the mechanism of action of *rec* genes on intragenic recombination (3.7). If it is assumed (Catcheside, 1966a) that *rec* genes control the activation or inactivation of the correction mechanism in the corresponding loci, a sudden change in *rf* would be observed only in crosses involving mutants characterized by a sharp inequality of $fk_{(a \to +)}$ and $fk_{(+ \to a)}$. However, such mutants are rarely observed (Gutz, 1968). Another explanation (Whitehouse, 1966; Smith, 1971) is that the rec^+ alleles inhibit proximal dissociation. However, the R_1 class is at best reduced by half, which completely fails to explain the sharp (by 10-30 times) decrease in *rf*.

According to the hypothesis now proposed the *rec* gene is the regulator of correction in a particular locus. It can either inactivate the correction mechanism or weaken the degree of directivity of correction. In either case the value of *rf* will be increased, and this increase is both allele-specific (Iha, 1969) and gene-specific. Choice between these alternatives can be made only by tetrad analysis. In the first case the presence of the *rec* allele leads to postmeiotic segregation, in the second case the frequency of appearance of t tetrads is sharply increased.

This explanation also is fully applicable to the findings of Ahmad et al. (1972). On the assumption that heterozygous inversion in one chromosome of *S. brevicolis* leads to increased correction in the *buff* gene, this can explain both the decrease in the frequency of postmeiotic segregation (3.8) and also the fact that $buff^+$ recombinants appear mainly in the R_1 class in crosses with inversion (52 of 58, compared with 29 of 57 without inversion).

The existence of a regulator gene assumes the presence of a specific site in the gene on which the repressor produced by the regulator. Gene specificity of repression of recombination (3.7) indicates that the operators of recombination (recombiners) resemble the operators of transcription in their properties. At the same time, it has been clearly shown in at least one case (Catcheside, 1968) that the operator and recombiner are identical. Repression of synthesis of the enzyme glutamate dehydrogenase, which is controlled by the *am* gene of *Neurospora crassa*, is independent of which allele of the *rec-3* gene is present in the given strain (consequently, *rec-3* is not the regulator of transcription).

Taking the analogy with the regulation of transcription further, the possibility of obtaining mutations in the recombiner can be predicted. The possibility cannot be ruled out that a mutation of this type has been found in *N. crassa* (page 95): *cog,* the recombiner of the *his-3* locus. In a cog^+ × *b* cog^+ and *a cog* × *b* cog^+ crosses the proximal polarity characteristic of this gene changes to distal. However, it is not yet clear why the character of the polarity is unchanged in *a cog*$^+$ × *b cog* crosses.

Heterozygosity in Bacteriophages. Correction models of recombination have been developed on the basis of facts obtained by tetrad analysis in fungi. Let us examine whether they provide a basis for the understanding of recombination mechanisms in other systems. Let us first look at the problem of heterozygosity in bacteriophages which, in the writer's opinion, is the key to the solution of the problem of the mechanism of recombination in these systems.

Phage heterozygotes were first discovered by Hershey and Chase (1951). They are formed in about equal frequencies (about 10^{-2} in phages T2, T4, and T1 and 10^{-4} in phage λ) when wild-type phages are crossed with any mutant phages. Phage heterozygotes are easily identified if the appropriate selective system is available, for during propagation they segregate to form a mixed clone consisting of particles of both genotypes, with the result that a sterile mottled plaque appears. Hershey and Chase found that heterozygotes formed in crosses between phages differing by two markers are rarely (in 3-5% of cases) heterozygous for both markers simultaneously, unless the markers are closely linked. The frequency of appearance of double heterozygotes in the case of closely linked markers is inversely proportional to the distance between them.

Several types of heterozygotes are formed in the progeny from crosses between two closely linked mutants (Table 27). Heterozygotes segregating during replication into the two original parental types are known as non-

TABLE 27. Genotypes and Frequencies of Appearance of Different Types of Progeny in *r 240* × *r 359* Crosses (from Edgar, 1958)

Conventional symbols		Genotype	Frequency in progeny
P_1 P_2	Parents	$r +$ $+ r$	0.5 0.5
R_1 R_2	Recombinants	$+ +$ $r\, r$	$7 \cdot 10^{-5}$ $7 \cdot 10^{-5}$
RH_1	Heterozygotes	$\frac{++}{r+}, \frac{++}{+r}$	$1 \cdot 10^{-5}$
RH_2		$\frac{rr}{r+}, \frac{rr}{+r}$	$1 \cdot 10^{-5}$
DRH		$\frac{++}{rr}$	?
NRH		$\frac{+r}{r+}$	$2 \cdot 10^{-2}$

recombinant (NRH). The reciprocal type, segregating two complementary recombinants, is called a double recombinant heterozygote (DRH). The remaining types are described as recombinant heterozygotes (RH), and they segregate into one of the parental types and one of the recombinant types. RH are heterozygous for only one of the markers participating in the cross. They account for a considerable proportion of the total number of recombinants formed in crosses of closely linked markers. In intragenic crosses they constitute 10-15% regardless of the distance between the sites (Edgar, 1958).

These results are easy to understand on the assumption that heterozygotes are formed in the process of recombination and that only a very small part of the phage genome can be in a heterozygous state. Hershey and Chase postulated that heterozygotes are essential intermediate structures in recombination and that their segregation leads to the formation of pure recombinants. Heterozygous particles observed in the progeny are those which are unable to segregate because they were removed from the pool of vegetative phage during maturation. This hypothesis has subsequently received support from a mass of experimental data. It was observed earlier (1.9) that the probability of crossing-over per map unit varies in different regions of the chromosome. In phage T4 a positive correlation was found between the probability of "saving" the marker in particular regions of the chromosome,* the frequency of genetic recombinations in these regions, and the frequency of formation of heterozygotes for these markers (Rottlander et al., 1967). For example, the frequency of formation of heterozygotes for gene 34 is twice as high as in the rII region (2% compared with 1%). The probability of saving marker 34 is also several times higher than that of saving marker rII. Other regions with an increased probability of heterozygote formation have been found near genes 25 and $h41$ (2 and > 3% respectively). Similar results were obtained in a study of phage T1, in which the increased frequency of recombinations at the ends of the chromosome also correlates well with the higher probabilities of saving the markers and the increased frequency of formation of heterozygotes for these markers (0.1-0.5% in the middle of the chromosome and up to 5% at its ends) (Michalke, 1967).

Hershey and Chase's idea that heterozygotes are intermediate structures in recombinant formation was confirmed by other investigators (Levinthal, 1954, 1959; Trautner, 1958; Kellenberger et al., 1962) who showed, by three-factor crosses of phages T2, T1, and λ that heterozygotes for the central marker are nearly always recombinant for the outside (flank) markers. This discovery could be interpreted as follows: in each act of recombination

---

*Saving the marker takes place as the result of genetic recombination between irradiated and unirradiated phages. The probability of saving is measured by the magnitude of the surviving fraction of wild-type phages irradiated with UV light and crossed with an unirradiated mutant.

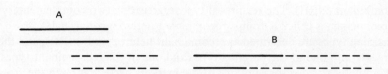

Fig. 90. Structure of phage heterozygotes: A) the diploid overlapping model; B) a molecular heterozygote. Continuous and broken lines denote DNA strands from different parents.

between two sites a heterozygous region is invariably formed as a result of partial overlapping of two parental fragments of chromosomes (Fig. 90). Approximate calculations (see below) show that the observed frequency of the heterozygotes in three-factor crosses of phage T4 is sufficient for this assumption. Of the two schemes proposed by Levinthal, the most popular is the second (Fig. 90B) according to which the genome of the heterozygous phage particle consists of an ordinary double-helical DNA molecule, and that heterozygosity arises because the complementary strands in the region of the junction of the two parental fragments carry different genetic information. Nevertheless the model of diploid overlapping (Fig. 90A) has also been supported by other workers (Edgar, 1958; Doermann and Boehner, 1963). Furthermore, facts completely inexplicable from the point of view of a scheme of true molecular heterozygosity (Fig. 90) have accumulated. Hertel (1963), for instance, found that one phage particle (T2) can be heterozygous for three alleles simultaneously. Nomura and Benzer (1961) found that deletions form fewer heterozygotes than point mutations (phage T4).

Streisinger et al. (1964) gave a completely unexpected explanation of these facts when they postulated that each chromosome of phage T2 (or T4) has a circular linkage map (*abcd....xyzabc*), since the ends are not fixed, because of inexact cutting of the chromosomes from the concatemers, any segment of the chromosome can be duplicated (*cd....xyzabcd* or *xyzab....xyz*). As a result of these circular permutations the genetic map becomes formally closed, although physically the chromosome is open-ended. Circular duplications are formed and lost as a result of recombination, which also leads to the appearance of phage heterozygotes. Streisinger postulated on the basis of Nomura and Benzer's results that two types of heterozygotes of phage T4 are formed: internal and circular. Point mutations can be detected in heterozygotes of both types, but deletions only in the circular type. Since heterozygotes of the last type are formed and lost during recombination, the probability of their appearance in the progeny depends on the rate of recombination, but not on the rate of replication. Internal heterozygotes are also formed during recombination but they disappear only during replication.

This difference served as the basis for an experiment which clearly demonstrated that two types of heterozygotes of phage T4 actually exist. When phages carrying point mutations in the *rII* region were crossed under conditions blocking DNA synthesis (FUDR) the frequency of heterozygotes in the progeny was increased five- to tenfold. If even one of the parental phages carried a deletion, the frequency of appearance of the heterozygotes was indistinguishable from the control (crossing without FUDR). In another elegant experiment Shalitin and Stahl (1965) showed that the proportion of RH among wild-type recombinants obtained from crossing mutants with point mutations in the *rII B* rose in the presence of FUDR from 16 to 63%, whereas the frequency of RH having a deletion was not increased at all (5.8% in both the control and the experiment).

Physicochemical evidence of the existence of circular terminal redundancies in T4 has also been obtained (Thomas, 1967). It has also been shown that chromosomes of heterozygous phages contain no structural changes in their DNA and that, consequently, internal heterozygotes have the structure represented in Fig. 90B (Anraku and Tomizawa, 1965). It has also been shown that heterozygotes of phage λ (the ends of the chromosome are fixed in this phage, so that terminal heterozygotes do not exist) are in fact formed as the result of breakages and reunions of the chromosomes (Meselson, 1965). The progeny obtained from crosses of heavy phage λ with light phage λ_c was studied in a density gradient. The distribution of phages of *c* type corresponds to the localization of the *c* locus on the map (Fig. 91). Phage

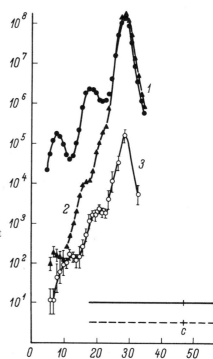

Fig. 91. Distribution in a CsCl density gradient of phage λ particles forming normal (1), clear (2), and mottled (3) sterile plaques (from Meselson, 1965). Cross ($C^{13}N^{15}$) λ × λ*c*. Position of the *c* locus on the map is shown in the bottom right-hand corner. Remainder of legend as in Figs. 57 and 58.

c/c^+ heterozygotes are found in peaks corresponding in density to 3/4 and 3/8 of the heavy parent. This means that their chromosomes consist of two different parental fragments joined together through a region of heterozygosity. It is curious that not all breakages in the c region are accompanied by the formation of heterozygotes (see below).

During recombination in phages molecular heterozygotes are thus formed. However, is the mechanism of their formation proposed by Levinthal (by end-to-end union of double-helical fragments) the only one possible? It was suggested on the basis of a study of the distribution of flank markers in heterozygotes of phage T4. At that time, however, nothing was known about terminal redundancies in this phage. Since terminal heterozygotes must always have an R configuration of the flanks,* their presence may distort the picture considerably. In fact, in Kellenberger's experiments on phage λ, fewer than 60% of RH were found (the remaining heterozygotes had a P configuration of the flank markers). To reduce the effect of circular heterozygotes to a minimum Berger (1965) conducted experiments on phage T4 in the presence of FUDR. In his experiments the frequency of appearance of RH was only 47%. In control crosses with point mutants the frequency of RH was somewhat higher (56%). In this way "anti-Levinthal" heterozygotes were found, i.e., heterozygotes with the P configuration of their flank markers. Honda and Uchida (1969) showed that exchanges between distant markers pass through a Levinthal structure, whereas exchanges between closely linked markers pass through an anti-Levinthal structure, so that high negative interference arises. In fact, PH can arise in the case of double crossing-over or repeated crossing in the pool. However, their frequency in that case would be considerably lower.

The excess of P configurations is explained (Berger, 1965; Doermann and Perma, 1967) by fitting the "traveling" DNA fragment into the unrepaired gap of the homologous molecule (Fig. 67). This explanation is exactly the same as that given by Watson et al. (5.2) for high negative interference. However, the most probable model of PH formation is the model of formation of the P configurations of flank markers in a standard flank system (5.4).

This mechanism does not lead to initial fragmentation of the parent chromosomes (as postulated by the schemes of Szybalski and Thomas; see also Dubinin et al., 1972), but to the formation of branched molecules as intermediate structures during recombination (Broker and Lehman, 1971). The frequency of double exchanges, according to Doermann and Parma (1967) is 90%, which agrees fully with the estimate of the relative frequency of semicrossing-over (page 178). It has been known for a long time that about 50% of the parental marker in phage T4 is transferred in large frag-

---

*In a cross between abc and the wild type $(a^+b^+c^+)$ b^+/b the terminal heterozygote must have two possible types of distribution of the a and c markers: $bc...a^+b^+$ or $b^+c^+...ab$, both recombinant for the a and c flank markers.

ments while the rest is transferred in segments amounting to about 1% of the length of the chromosome. The former evidently correspond to fragmentation of DNA into double-helical segments during crossing over, the latter to the formation of single-stranded fragments during semicrossing-over.

Does correction take place in phages? If the phage heterozygotes were not corrected their frequency for the central marker in a three-point cross would be high enough. It can be calculated if the distance between the markers, the length of the heterozygous region,† and the number of generations in the pool (m) are known. In λ m_6 c mi $\times$ $\lambda+$ crosses, 2.6×10^{-3} c/c^+ heterozygotes ought to appear, since $d_{m_6\text{-}mi}$ = 15 map units, x = 2 map units, and m = 5.‡ The observed frequency of appearance of c/c^+ heterozygotes is 1×10^{-4} (Kellenberger et al., 1962), i.e., it is 25 times smaller. The frequency of heterozygotes in T-even phages is 1×10^{-2}, which was apparently sufficient (Levinthal, 1959). However, Levinthal's calculations were based on the assumption that all heterozygotes observed are internal. The size of the heterozygous region was not precisely known. Calculations based on the results obtained by Berger, who used only point mutations in FUDR crosses, also show a deficiency of heterozygotes. In multiple-factor crosses involving *rII* mutants of phage T4 17% of heterozygotes for central markers were found. The flank markers *r2* and *r65* were at a distance of 4 map units apart. The length of the heterozygous region was 2-3 map units, m = 0. Between 50 and 75% of heterozygotes ought therefore to have been observed. These calculations apply to crosses involving point mutations. In general no deletions are found in internal heterozygotes. Berger found that in crosses involving a deletion and several point mutations in the presence of FUDR the probability of detection of point mutations closely linked with the deletion also were sharply reduced. Drake (1966) observed that the frequency of appearance of heterozygotes for point mutations induced by acridine is heterogeneous. If the mutation behaves like a small deletion during crossing, it is less likely to be detected than an *amber* mutation (simple replacement). Consequently, either deletions cannot be incorporated into internal heterozygotes or they are effectively corrected after incorporation. The second suggestion is most likely for deletion heterozygotes in phage λ have recently been found with the electron microscope: they form characteristic loops (Davis, 1968; Westmoreland et al., 1969).

†The length of the heterozygous region (x) can be determined in two ways: 1) $t = \dfrac{d\,(n+2)}{n}$, where $x > d$, $n = \dfrac{RH}{NRH}$ and d is the distance in map units; the value of x determined in this way for phage λ is 1.4-2.1 map units (Kellenberger et al., 1962); 2) by determining the mean number of markers incorporated into the heterozygous region in a multiple-factor cross; the true size of the region in phage T4 can be determined only in crosses involving point mutations or in the presence of FUDR. In either case x = 2-3 map units (Berger, 1965).

‡ $fH = \dfrac{x}{md}$.

The fact that high negative interference is also observed in three-point crosses involving deletions (Doermann and Parma, 1967; Matvienko, 1969, 1972a,b) is also evidence that the formation of molecular heterozygotes and subsequent correction take place (see page 222).

Benz and Berger (1973) determined the ratio of [wt/r(progeny)] to [wt/r(parental)] in FUDR crosses: for 9 point mutations it was 1.2, for deletions 0.8, and for mutations of the insertion type it was also greater than 1. The impression was obtained that the excess of DNA is always excised. It was shown that the excision process is under the control of the x and v genes, whereas gene 43 and the host genes $polA$, $EndoI$, $recA$, $recB$, and $uvrA$ have no effect on it.

Clear and unambiguous evidence of correction of phage heterozygotes has also been obtained recently for phage SPP1 of *Bacillus subtilis* (Spatz and Trautner, 1970). These experiments also demonstrated the directivity of correction.

The complementary strands of the DNA of this phage separate readily in a CsCl density gradient. After preparative separation of the strands they were treated with hydroxylamine and mutations influencing the type of sterile spot. Since hydroxylamine converts G–C into A–T, the orientation of the original base pairs relative to the heavy (H) and light (L) strands can be determined. Heterozygous molecules consisting of a wild-type H strand (+) and a mutant (−) L strand, or vice versa (H^-/L^+), were then constructed. Cells of *Bacillus subtilis* were infected with these heterozygotes and the phage yield from the individual cells analyzed. The following figures were obtained for mutation 161. DNA H^+/L^-: 69% of the 876 cells liberated only wild-type phages, 8% only mutant phages, and 23% a mixed progeny. DNA H^-/L^+: 5% of the 805 cells liberated wild-type phages, 80% mutant, and 15% phages of both types. Bases thus were excised mainly from the light strand (regardless of the allelic information it carried). Presumably the correcting enzyme specifically excises purines (A and G), for the light chain in the first case (H^+/L^-) contained G, while in the second case (H^-/L^+) it contained A. However, analysis of other mutations showed that this is not the rule, and in other cases pyrimidines are chiefly excised. These workers concluded that the factor determining the direction of correction is the composition of the bases close to the mutation.

Evidence in support of the correction hypothesis has been obtained by transfection with heteroduplexes of phage ϕX 174 (Baas and Jansz, 1972), and transformation of phage T4B (Vinetskii and Aleksandrova, 1972).

After the discovery of the three recombination systems operating on the chromosome of phage λ it became clear that the classical experiments of Meselson and Kellenberger would have to be reproduced for each of them separately. Moreover, since vegetative replication of the phages was not blocked in these early experiments it was impossible both to determine the precise quantity of reparative synthesis and to discriminate between cross-

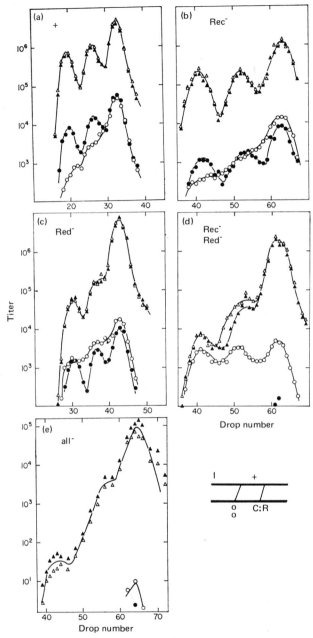

Fig. 92. The density distributions of i^λ progenies from the density-labeled repressor cross $susJ6 \times cI857\ susR5$ (from Stahl et al., 1972a). a) Red^+Int^+ phage crossed in the Rec^+ host $QR47$; b) Red^+Int^+ phage crossed in the Rec^- host $QR48$; c) Red^-Int^+ phage crossed in $QR47$; d) Red^-Int^+ phage crossed in $QR48$; e) Red^-Int^- phage crossed in $QR48$. △ total c^+ phage, ▲ total c phage, ○ $c^+\ sus^+$ recombinants, ● $c\ sus^+$ recombinants.

over and semicrossover transfer of the label. Two systems with partial (Stahl and Stahl, 1971a,b) and complete (McMilin and Russo, 1972) blocking of DNA synthesis have recently been developed. In the first system, under both Rec$^+$Red$^+$ and Rec$^-$ conditions when two heavy (H) parents marked by the J, cI, and R genes were crossed no recombinants for $J-cI$ were contained in the H/H peak (Fig. 92a,b). In $H \times L$ crosses with complete block of DNA synthesis only those recombinants whose appearance requires transfer near cI or between cI and R (Rec$^-$ conditions) can be detected (Stahl et al., 1972a,b, 1973). Under Red$^-$ conditions (Stahl et al., 1973) a uniform distribution of recombinants is observed along the chromosome. These facts led to the conclusion that Red-induced recombination in the central part of the chromosome is accompanied by widespread DNA synthesis, and if this is inhibited the number of recombinants in this region is reduced by 20 times compared with the number of recombinants in the $cI-R$ region ($EJ/cR = 20$; Stahl et al., 1972b). The quantity of DNA synthesis connected with recombination is estimated to be 10-50% of the total phage chromosome. At the same time, in experiments on a system with complete blocking of replication in $H \times H$ crosses only 2-5% of synthesis was found for recombination both in the center and in the $cI-R$ region (McMilin and Russo, 1972; Russo, 1973), and this was true under both Rec$^+$ and Rec$^-$ conditions.

Furthermore, in a system with incomplete block inactivation of the Red system in $H \times H$ crosses leads to a sharp decrease in the height of the parental H/H peak relative to the L/L peak (Fig. 92c), whereas inactivation of the Rec system does not affect its height (Fig. 92b), although under Rec$^-$Red$^-$ conditions it becomes lower still (Fig. 92d). The H/H peak disappears almost completely if all three recombination system are inactivated (Fig. 92e). This shows that recombination is absolutely necessary for maturation of the unreplicated H/H particles (and for some of the H/L). How can this condition be understood?

After injection into the cell the linear phage chromosome is closed into a ring by means of its sticky ends (AR joint) and replicates several times as the θ form, the products of which also are rings. This is followed by a switch to the stage of σ replication of the rotating wheel type, resulting in the formation of linear concatemers ("tails"), which mature: they are cut by the Ter protein of gene A at the AR joint and encapsulated. Evidently only the chromosome enclosed between two AR joints can mature. If DNA synthesis is blocked completely this condition can be satisfied only if crossing-over takes place between two rings with the formation of a figure eight. Folkmanis and Freifelder (cited by Stahl et al., 1972a) showed that each of the recombination systems working separately can convert circular λ monomers into dimers in the absence of DNA synthesis. Nevertheless, Stahl's group (Stahl et al., 1973) considers that the Red system cannot carry out a reciprocal exchange and, because of this, it acts as a

§5.6] THEORY OF DIRECTED CORRECTION 209

switch from the θ to the σ replication. If recombination under the influence of the Red system takes place in the $cI-R$ region, for the wheel to untwist and the recombinant to emerge into the tail a very small quantity of synthesis is necessary, but if the recombination takes place in the middle of the chromosome extensive DNA synthesis is required for maturation of the recombinant. The Rec system can give a figure eight at any point of the chromosome.

This model is contradicted by the findings of Enquist and Skalka (1973), as a result of which the picture must be modified somewhat. The $\theta \to \sigma$ switch is probably facilitated in the presence of any of these systems, although it can take place in their absence. The distinguishing feature of these systems is that in the λ exonuclease there is no double-helical exonucleotic activity, which is present in exonuclease V. Exonuclease V therefore destroys the tails and the switch to σ replication can occur only after sufficient protein of gene γ, inhibiting RecBC nuclease, has been synthesized. It also follows from the findings of Enquist and Skalka that the Red system is capable of crossing-over, for $\lambda \gamma^-$ replicates in Rec$^-$ as multimeric rings, while $\lambda \gamma^- red^-$ under the same conditions exists entirely in the form of monomeric rings, which mature very badly (feckless). Asymmetry of the recombination events may thus be connected with the induction of Red recombination in the $cI-OP$ region when DNA synthesis is blocked.

Furthermore, it is impossible to determine the precise quantity of DNA synthesis associated with crossing-over since the H/H parental peak is held at the crossover if synthesis is blocked completely. For this reason, displacement of the recombinant peaks in experiments of this type must be ascribed to the presence of correction synthesis of DNA. However, the quantity of reparative synthesis in crossing-over evidently cannot be large, for the H/H parental peak is not displaced very considerably relative to the L/L reference peak.

One further conclusion can be drawn from the foregoing facts. If synthesis is blocked completely, semicrossover transfer of the label cannot be detected. Meanwhile, in the experiments of Russo (1973), under conditions in which the recombinants were bound to inherit 18.5% of the label of the heavy parent, only 10% transfer was found under both Rec$^+$ and Rec$^-$ conditions. Probably the region of overlapping during crossing-over may cover up to one-fifth of the entire chromosome of phage λ.

If the Red system is capable of reciprocal crossing-over, the nonreciprocity of recombination between genes in phages may thus be entirely connected with conversion. This is proved by elegant experiments on the defective phage λdv (Kellenberger, 1971; Berg, 1971), which showed that rescue of the $susP^+$ marker from this phage can be achieved in two ways: either by insertion of a λdv ring chromosome into the chromosome of the rescuer phage (λb2b5), with the resulting formation of phages with increased density (D peak), or the $susP^+$ marker appears in the S peak with the density of the rescuer phage. The frequency of these cases was found to be 25% by Berg and more than 50% by Kellenberger. All the $susP^+$ phages from the S peak were homozygous, while those from the D peak were heterozygous. The picture was identical both in the Rec⁻ bacteria and in those with Red⁻ phages. These workers draw several important conclusions from these experiments: 1) Red and Rec systems are capable of reciprocal crossing-over, 2) about half of the recombinations take place by semicrossing-over (the $susP^+$ recombinants in the S peak), and 3) correction accompanies semicrossing-over.

Let us now examine the crossing experiments of Weil (1969). The distance between $susA$ and $b2$ corresponds to about half of the chromosome. The length of the heteroduplex overlap is one-tenth of the chromosome. Consequently, the probability of conversion of any marker is 0.4 (0.1 × 2/0.5). The coefficient of correlation between reciprocal events in this region should be 0.6. In fact, $r = 0.5$ (see page 33). The closer together the markers, the smaller the value of r should be. In fact, r was much smaller in Kaiser's (1955) experiments (the segment of the chromosome was half as large and was twice as long as rf, since the crossing was carried out on UV-irradiated bacteria), and reciprocal recombinants are virtually completely absent in transfection with $ARar$ λ dimers (Melechen and Hudnik-Plevnik, 1972).

Mechanisms of Recombination in Merozygous Systems. During transformation of bacteria only one (either) of the strands of the donor DNA molecule is built up into the chromosome of the recipient cell. Consequently, the mechanism of recombination (both within the gene and between genes) during transformation is by semicrossing-over. This conclusion has been reached as a result of genetic and physicochemical research.

Marmur and co-workers (Doty et al., 1960; Marmur and Lane, 1960) developed a method of obtaining hybrid (see second footnote on page 159) DNA molecules *in vitro*. This method consists of thermal denaturation of a mixture of two DNA preparations obtained from strains differing in one or more allele markers. Subsequent slow cooling of the mixture ("annealing") leads to renaturation of the DNA, during which strands from different molecules are joined (the process is described as hybridization of DNA). If transforming DNA is hybridized with a DNA which is homologous but genetically inert with respect to the chosen marker, the proportion of molecules carrying the active markers must be increased. In the limiting case the number of these molecules is doubled as the result of hybridization. This can be achieved by hybridization in the presence of a large excess of inert DNA, when it is more probable that molecules in which only one of the complementary strands carries the active marker will be formed during renaturation (molecular heterozygotes).

If double-helical molecules of donor DNA are incorporated into the cell chromosome during transformation the molecular heterozygotes will be just as effective in transformation experiments as ordinary molecules. Consequently, in this case hybridization must increase the effectiveness of transformation. In one of a series of experiments (Bresler et al., 1964a,b) aimed at finding this effect, DNA isolated from a prototrophic strain of *Bacillus subtilis* SHgw was hybridized with a DNA preparation obtained from strain 168 *ind*⁻ (indole deficient). The ratio between the amounts of SHgw DNA and *ind*⁻ DNA on hybridization was 1 : 10. With this ratio between the components after denaturation and renaturation 90% of the strands of SHgw DNA ought to go into hybrid molecules. A mixture of the two DNA preparations in the same ratio, after denaturation and annealing separately, was used as the control. Transformation was carried out on *B. subtilis* strain SB_{25} (*ind*⁻*his*⁻). A marker for which both strands of hybrid molecules were active (Table 28) was used as an indicator of the level of transformation in each tube. The levels of transformation (LT) for the heterozygous locus in hybrid DNA (LT_h) and for the indicator loci (LT_i) in the hybrid preparation and in the mixture were determined. If the efficiency of DNA hybridization was higher, the following inequality should apply:

$$K = \frac{(LT_h/LT_i)_{hybrid}}{(LT_h/LT_i)_{mixture}} > 1.$$

The results (Table 28) show that, despite an increase in the number of molecules carrying the active marker, the transforming activity of the preparation is not increased. Since the expected effect did not arise it was postulated that usually only one of the strands of the donor DNA molecule is

TABLE 28. Effectiveness of Hybrid DNA Preparations in Transformation

Original mixture 1 *ind*⁺ : 10 *ind*⁻	Hybrid molecules	Genotype of recipient	K
$\overline{\begin{array}{cc}ind^- & his^+\\ ind^- & his^+\end{array}}$	$\overline{\begin{array}{cc}ind^- & his^+\\ ind^+ & his^+\end{array}}$		
		$\overline{\begin{array}{cc}ind^- & his^-\\ ind^- & his^-\end{array}}$	1.04 ± 0.04
$\overline{\begin{array}{cc}ind^+ & his^+\\ ind^+ & his^+\end{array}}$	$\overline{\begin{array}{cc}ind^+ & his^+\\ ind^- & his^+\end{array}}$		

incorporated into the chromosome of the recipient cell. Moreover, the extremely low frequency of appearance of mixed clones during transformation of a double auxotrophic strain by heterozygous DNA molecules also pointed to this possibility (see also the analogous experiments using separated DNA strands; Strauss, 1970). Direct physicochemical proof of the mechanism of single-stranded incorporation during transformation was soon obtained (Bodmer and Ganesan, 1964; Fox and Allen, 1964; Fox, 1966; Goodgal and Postel, 1967).

In a typical experiment light recipient bacteria were transformed with heavy radioactive DNA. At specified time intervals after contact with DNA the cells were analyzed and the DNA isolated from them was centrifuged in a density gradient. Biological activity of the donor DNA and radioactive label were found in the peak corresponding in density to recipient (light) DNA. If the DNA isolated from this peak was degraded by hydrodynamic forces or by sonication so that its molecular weight fell from $(20\text{-}40) \times 10^6$ to $(5\text{-}6) \times 10^6$ daltons, activity passed into the hybrid zone. Denaturation of the material led to disappearance of the hybrid zone.

The use of genetic and physicochemical methods thus showed that only one of the complementary strands of the donor DNA molecule is incorporated into the chromosome of the recipient cell during transformation. It accordingly becomes clear that the transformation method offers a unique opportunity for the solution of the correction problem, for heterozygosity for any marker can be reliably obtained in the chromosome of the recipient cell, and the subsequent fate of this chromosome can then be studied.

If correction does not take place, after semiconservative replication the heterozygote must segregate. Consequently, the transformants will always form mixed clones in which, side by side with transformed cells, there will be an approximately equal number of cells with recipient genotype. If correction of the molecular heterozygotes does occur, pure clones of transformants will be observed with a definite probability. To study the possibility of correction of molecular heterozygotes the method of clonal analysis was used (Bresler et al., 1968; Kushev et al., 1970). Auxotrophic mutants of *B. subtilis* were treated with DNA isolated from wild-type strains (SHgw) and seeded on nonselective medium, allowing both recipient cells and recombinants to propagate equally effectively. By means of Lederberg's impression method the growing colonies were transferred from these dishes to dishes with minimal medium so that colonies of recombinants capable of propagating on it could be identified. The recombinant colonies were analyzed to discover if they contained cells of the recipient genotype. The results of these experiments are given in Table 29. Since the frequency of appearance of pure clones differs sharply for two closely linked markers (*thy B* and *met*), the correction hypothesis is the only one which can satisfactorily explain the results. These experiments are convincing evidence in

TABLE 29. Frequency of Appearance of Pure Clones among Transformants for Individual Markers

Marker	Character	Total number of transform-ant clones	Number of pure clones among them	Number of pure clones (p^+), in %
met	Requirement of methionine	259	47	18 ± 2
ind	Requirement of indole or tryptophan	235	86	37 ± 3
trp_3	Requirement of tryptophan	141	65	46 ± 4
$thy\ B$	Requirement of thymine	107	78	72 ± 4
$rib^c\text{-}1$	Constitutive synthesis of riboflavine	85	67	80 ± 4

support of the correction of molecular heterozygosity postulated by Holliday. The differences between the frequencies of appearance of the pure clones mean that some markers are readily corrected while others are not, i.e., that correction is allele-specific.

If sites of heterozygosity in the cell are corrected independently, double transformants ought to form clones of the following types with a definite probability (Table 30). Since p^- in experiments of this type cannot be detected, as a first approximation it can be considered that $p^0 + p^+ = 1$ (page 215).

The appearance of clones mixed for only one of the markers ($a^{+/-}b^+$ or $a^+b^{+/-}$) is decisive for reliable conclusions to be drawn regarding the dependence or independence of correction of the two sites of heterozygosity (in the case of independent correction of two sites of heterozygosity the frequencies of their appearance were determined by the product of the probabilities $p_a^0 p_b^+$ and $p_a^+ p_b^0$). Their appearance can easily be detected in experiments with primary selection for one character. Experiments were

TABLE 30. Appearance of Clones of Different Types during Transformation of Double Auxotrophic Strain of *B. subtilis* by SHgw DNA

Parameters of correction	p_a^+	p_a^0
p_b^+	$a^+ b^+$	$a^{+/-} b^+$
p_b^0	$a^+ b^{+/-}$	$a^{+/-} b^{+/-}$

Note. p_a^+ and p_b^+ denote probabilities of correction of molecular heterozygotes for markers a and b.

TABLE 31. Analysis of Clones of Double Transformants Grown on Partially Selective Media

Selected character	Unselected character	Number of clones		Theoretically expected, in %	Total
		Mixed for the unselected character			
		abs.	in %		
ind	tyr	11	2 ± 0.6	20	500
thy B	met	40	24 ± 3.0	56	165
ind	thy B	17	12 ± 3.0	15	140
met	ind	29	16 ± 4.0	12	183
ind	met	80	40 ± 5.0	32	199
trp_3	rib^c-1	15	7 ± 2.0	8	219

Note. The theoretically expected percentage was calculated on the assumption that correction is independent. The values of p^0 and p^+ for each marker were determined in experiments without preliminary selection (see Table 29).

carried out as follows. After transformation of double auxotrophic strains with DNA of wild type the cell population was grown on a partially selective medium on which only recombinants for one or both markers could propagate. Colonies growing on partially selective medium and transferred by the impression method to selective medium were analyzed for mixing for the unselected marker. The results are given in Table 31.

The following consideration is vital to the analysis of these results. The *ind* and *tyr* markers are linked, the distance between them is of the order of 30 map units, the distance between *thy B* and *met* is more than 50 map units, and the pairs *ind–met*, *rib^c-1–trp_3*, and *ind–thy B* are not linked. The theoretically expected frequencies were obtained only in the case of unlinked markers. Consequently, if markers are close together they are corrected simultaneously (linked correction).

Using the method of preliminary selection, Louarn and Sicard (1968a,b, 1969) showed that correction also takes place at the *ami A* locus of *Pneumococcus,* and they also found correlation between the integration efficiency of a marker and the degree of its correction (markers with high integration efficiency were corrected to a lesser degree than markers with low efficiency). Meanwhile Guerrini and Fox (Fox, 1966; Guerrini and Fox, 1968a,b) did not find pure transformants for *ad* markers in *Pneumococcus* in experiments on unselective medium (without selection). Pure transformants appeared only after treatment of the population with UV irradiation or mitomycin C. Fox interpreted these results on the basis of a hypothesis of incidental homozygosis (see page 224). However, alternative explanations are possible: (1) uncorrectable markers were used, or (2) UV irradiation eliminated the multinuclear cells which were unavoidably present despite preceding sonication and heat treatment. The possibility of obtaining double transformants under the influence of trans-diheterozygous DNA molecules is further evidence in support of a correction mechanism (Bresler et al., 1964c; Roger, 1972).

In the experiments considered above only two p parameters were recorded: p^+ and p^0.* The probability of correction to the recipient text (in experiments of this type p^-) cannot be directly measured. Only relative values of p^- can be determined by measuring the integration efficiency (IE) of the corresponding markers.

The idea of correction as an explanation of differences in IE of the markers during transformation was first involed by Ephrussi-Taylor and Gray (1966). Previously attempts had been made to associate IE with the position of the marker on the molecule (Sicard and Ephrussi-Taylor, 1965), with the size of the mutation injury (Ephrussi-Taylor, 1961), and with the probability of pairing (Lacks, 1966). This last hypothesis was based on the assumption that certain types of mutation injuries (particularly deletions) disturb complementarity so severely that the probability of formation of a molecular heterozygote is considerably reduce. Against this hypothesis is the fact that the IE of many (even large) deletions does not differ from unity. Direct proof that correction to the donor text (p^+) is possible was obtained in the experiments on clonal analysis described above. It can evidently now be regarded as proved that IE does not depend on the probability of synapsis of donor DNA with the chromosome, but on the probability of its removal from the chromosome by a process of correction. Recently (Lacks, 1970) *hex* mutants of *Pneumococcus* with injury to their correction system have been isolated. During the transformation of these strains all markers are integrated with high efficiency.

Since we do not know p^- for the reference marker whose IE is taken as 1, the determined values of p^- are relative. The absolute IE can be determined only if p^- of the reference marker is 0 (when the IE of this marker is strictly equal to 1). In the case of LE markers (2.12) correction to the recipient text (p^-) thus takes place on the average ten times more often than the two other possible events: absence of correction (p^0) and correction to the donor (p^+), under the condition that $p^+ + p^0 + p^- = 1$. We have already seen (2.12) the effect of differences between IE of the markers during mapping.

*It is evident that neither of these parameters is a true probability because the third parameter p^- is unknown. By basing the calculations on the number of pure and mixed clones, instead of the probabilities p^+ and p^0, satisfying the usual standard rule $p^+ + p^- + p^0 = 1$, other values were determined:

$$\frac{p^+}{p^+ + p^0}$$

and

$$\frac{p^0}{p^0 + p^+} = 1 - \frac{p^+}{p^+ + p^0}.$$

This had no effect on determination of the probability of independent correction of the two markers because the values of p^+ and p^0 obtained are proportional to the true probabilities.

Genetic maps can be drawn with sufficient determinancy only if the markers have identical IE. Otherwise it is impossible to determine IE in separate experiments and to introduce appropriate corrections for IE during mapping. For example, IE_{his} for *B. subtilis* is only half the value of IE_{ind} and IE_{tyr} (personal observations). This explains the difference between *rf* values in reciprocal crosses (Anagnostopoulos and Crawford, 1961): his_2 (H25) → *ind* (168) = 45%; 168 → H25 = 26%. In this case *rf* was defined as the ratio of the number of ind^+his^+ to the number of $ind^+his^+ + ind^+his^+$ forms (in the first cross) and to the number of $ind^+his^+ + ind^+his^-$ forms (in the second cross). Consequently (1.17), we can write

$$rf_{his-ind} = \frac{rf_1 rf_2 (1-rf_3)}{rf_1 (1-rf_2) rf_3 + rf_1 rf_2 (1-rf_3)}$$

and

$$rf_{ind-his} = \frac{(1-rf_1) rf_2 rf_3}{rf_1 (1-rf_2) rf_3 + (1-rf_1) rf_2 rf_3}.$$

Taking $IE_{ind} = a$ and $IE_{his} = b$, where $b = 0.5a$, and bearing in mind the fact that IE of two markers simultaneously is determined by IE of the LE marker, we can rewrite the equations as follows:

$$rf_{his-ind} = \frac{arf_1 rf_2 (1-rf_3)}{brf_1 (1-rf_2) rf_3 + arf_1 rf_2 (1-rf_3)} \quad (1)$$

and

$$rf_{ind-his} = \frac{b(1-rf_1) rf_2 rf_3}{brf_1 (1-rf_2) rf_3 + brf_1 rf_2 (1-rf_3)}. \quad (2)$$

In the latter case the true value of *rf* is obtained (*b* cancels out). (In the absence of information on IE of the markers, reciprocal crosses must be carried out and values of rf_{min} used for mapping.) Consequently, differences between IE of the markers may lead to an incraease in *rf* up to values of $rf > 0.5$ (see, for example, Sicard and Ephrussi-Taylor, 1965). The importance of linked correction can well be emphasized at this point. If *his* and *ind* markers were corrected independently, the *rf* values in reciprocal crosses would be equal. The equation for $rf_{his-ind}$ could be written in the form

$$rf_{his-ind} = \frac{arf_1 rf_2 (1-rf_3)}{arf_1 (1-rf_2) rf_3 + arf_1 rf_2 (1-rf_3)},$$

where *a* appears in both the numerator and the denominator, just as *b* did in the equaltion for $rf_{ind-his}$.

Hence,

$$\frac{rf_{his-ind}}{rf_{ind-his}} = \frac{[b\, rf_1 (1-rf_2) rf_3 + arf_1 rf_2 (1-rf_3)] b(1-rf_1) rf_2 rf_3}{arf_1 rf_2 (1-rf_3) [brf_1 (1-rf_2) rf_3 + b(1-rf_1) rf_2 rf_3]}. \quad (3)$$

The values of rf_1 and rf_3 are close to the limiting value of 0.5 (see below) and the true genetic distance between *ind* and *his* is 0.25 [from Equation (2)]. Substituting these values in Equation (3), we obtain

$$\frac{rf_{his-ind}}{rf_{ind-his}} = 1.6,$$

in full agreement with the experimental findings.

It was shown earlier (1.17) that during conjugation the *rf* values determined in two-marker crosses can be regarded as absolute (provided that the appropriate conditions are satisfied). Can the values of *rf* during transformation also be regarded as absolute? Using Bailey's method (1.17) and the results obtained by V. L. Kalinin (personal communication, 1970), let us calculate the values of rf_1 and rf_4, where rf_1 and rf_4 are the distances from *ind* to the left end of the exogenote and from *tyr* to its right end respectively, and $rf_2 = rf_{ind\text{-}tyr}$ and $rf_3 = rf_{his\text{-}tyr}$ (the order of the markers is *ind, his, tyr*).

In the case of preliminary selection for the terminal markers *ind* and *tyr*, the analysis has to consider four classes of recombinants (Table 32), for which the following equations can be written:

Selection for ind^+
$$\begin{cases} ind^+his^-tyr^- = rf_1 rf_2 (1-rf_3)(1-rf_4) \\ ind^+his^+tyr^- = rf_1(1-rf_2) rf_3 (1-rf_4) \\ ind^+his^-tyr^+ = rf_1 rf_2 rf_3 rf_4 \\ ind^+his^+tyr^+ = rf_1(1-rf_2)(1-rf_3) rf_4 \\ ind^-his^+tyr^+ = (1-rf_1) rf_2 (1-rf_3) rf_4 \\ ind^-his^-tyr^+ = (1-rf_1)(1-rf_2) rf_3 rf_4 \end{cases}$$
Selection for tyr^+

TABLE 32. Number of Transformants of Different Classes in $ind^+his^-tyr^+ \rightarrow ind^-his^-tyr^-$ Crosses (from data of V. L. Kalinin)

Marker for which preliminary selection carried out	Class of transformants and their numbers			
	ind⁻his⁻tyr⁺	ind⁻his⁺tyr⁻	ind⁻his⁺tyr⁺	ind⁺his⁻tyr⁻
ind⁺	—	—	—	555
his⁺	—	106	610	—
tyr⁺	250	—	713	—

Marker for which preliminary selection carried out	Class of transformants and their numbers		
	ind⁺his⁻tyr⁺	ind⁺his⁺tyr⁻	ind⁺his⁺tyr⁺
ind⁺	40	355	1826
his⁺	—	325	1870
tyr⁺	46	—	1845

Using the equations in Section 1.17, we next determine rf_2, rf_3, and rf_4 from the results obtained by selection for ind^+, and the values of rf_1, rf_2, and rf_3 from the results obtained by selection for tyr^+. Comparison of the two sets of values calculated for rf_2 (0.16-0.21) demonstrates the reliability of the method. The values of rf_1 and rf_4 approximate to 0.5 ($rf_1 = 0.41$ and $rf_4 = 0.38$). However, as was shown in 1.17, if the ends of the exogenote equal 0.5, the rf values determined in two-point crosses are absolute.

Results obtained by E. I. Cherepenko (personal communication, 1970) when mapping mutations in the riboflavin operon of *B. subtilis* show that the values of rf_1 and rf_4 are independent of the value of $rf_2 + rf_3$.

In transformation experiments a very high probability of recombination is observed. The tryptophan synthetase gene of *B. subtilis* measures 10 map units (Carlton, 1966). This means that the physical scale of the map is 100 nucleotides per map unit. The scale is the same for transduction in *Escherichia coli* (Yanofsky et al., 1964). Yet during conjugation the scale of the map is 25 times smaller. What is responsible for these differences? A possible explanation could be that during conjugation recombination takes place at the level of double-helical DNA molecules (crossing-over), while during transduction and transformation it takes place at the level of single strands (semicrossing-over).

However, this explanation will not work because it has been shown that during transduction with phage P22 recombination takes place at the level of the double helix (Ebel-Tsipis et al., 1972). In this case the presence of high negative interference can be understood, but the problem of how the even exchanges take place requires consideration (see below). The concrete mechanism of recombination during transformation is evidently as follows. Long chains of donor DNA (80×10^6 to 90×10^6 daltons) attach themselves to the cell surface and as they penetrate into the periplasmic space they are cut by endonuclease into double-helical fragments (DSF) with molecular weights of 8×10^6 to 9×10^6 daltons (Dubnau and Cirigliano, 1972a,b). They then (possibly from one particular end, because the transfer of genetic markers exhibits polarity) penetrate through the cell membrane, conjugate [after partial (Goodgal and Notani, 1968) or complete denaturation] with the recipient chromosome, and one (either) strand breaks up into single-stranded fragments (SSF) with a molecular weight of 1.2×10^5 daltons and acid-soluble products. By this behavior *B. subtilis* and *H. influenzae* differ from pneumococci in which the breaking up of one of the strands takes place at the time of penetration of the molecule into the cell (Lacks, 1966). The donor DNA molecule of *B. subtilis* is introduced into the recipient chromosome with an efficiency of 64% per strand. This means that one DNA strand with a molecular weight of 25×10^6 daltons is inserted into one region of the chromosome. However, it is not inserted as a single strand, but as several discrete segments with a mean molecular weight of 2.8

$\times 10^6$, or of 6×10^6 if calculated for the double helix. This value agrees very closely with the estimate of the physical scale of the genetic map obtained above on the basis of Carlton's data (Dubnau and Cirigliano, 1972c). For this reason very large deletions cannot be transformed to the wild type (Adams, 1972). During integration of the DSF, SSF with a total molecular weight of 3×10^6 (for the double helix) are eliminated from it, evidently as a result of recombination at the ends, as may be deduced from the mathematical model of recombination during transformation (see above and also Cato and Guild, 1968). DNA with a molecular weight below 3×10^6 daltons thus does not possess transforming activity (Morrison and Guild, 1972).

What is the concrete mechanism of interaction between the DNA molecules during integration (Fig. 92)? Information on the state of the chromosome of the competent cell is very important to the understanding of this problem. To begin with, a local fluctuation denaturation may take place in a segment consisting of 6 to 8 nucleotide pairs, as is postulated for pneumococci (Collins and Guild, 1972; Shoemaker and Guild, 1972). This is followed by conjugation, rupture, and exonucleotic assimilation of the donor strand. In this case degradation of the recipient DNA will be induced by recombination, as occurs in *H. influenzae* (Steinhart and Herriott, 1968). The process of expulsion of the host DNA takes place through the action of homologous transforming DNA and depends on the functions of the *rec1* and *rec2* genes (Beattie and Setlow, 1971). The displaced host material is not reciprocally joined to the donor material but is degraded.

In *B. subtilis* (Harris and Barr, 1969, 1971a,b; Kohoutova et al., 1970) DNA liberated from the competent cells changes the buoyant density in CsCl by 0.03 g/ml during a change in pH from 8 to 11. These workers incline to the view that about 5% of the entire genome of the competent cell is in a single-stranded state and DNA molecules, penetrating into the cell simply adhere to these segments several thousand nucleotides long. Such a mechanism seems very doubtful, for the possibility of multiple recombination events in these regions must be borne in mind.

The result of integration is a heteroduplex, supported only by hydrogen bonds (Arwert and Venema, 1973), after which (in rec^+) covalent bond are formed (Davidoff-Abelson and Dubnau, 1971; Zadražil and Fučik, 1971). In a pneumococcal mutant which has lost its ATP-dependent nuclease and which transforms approximately 6 times less readily than the wild type, the

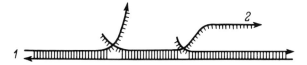

Fig. 93. Mechanism of recombination during bacterial transformation: 1) chromosome, 2) strand of donor DNA molecule.

donor–recipient complex is formed in normal amounts. This enzyme is perhaps necessary for the later stages of integration, e.g., for removal of the excess DNA from the complex (Vovis, 1973; see the survey by Notani and Setlow, 1973).

Furthermore, if the site of heterozygosity is in the hybrid region, the correction mechanism is activated. During linked correction several thousands of nucleotides can be eliminated. This is a large number, but should not be surprising. The Rec system can cut out and replace up to 3000 nucleotides (Cooper and Hanawalt, 1972a,b). The shift of the H/H recombinant peak in the phage (McMilin and Russo, 1972) also enables the scale of DNA extrasynthesis of this same order to be determined. Reparative synthesis of DNA over such long distances can easily be recorded provided that the experimental conditions are suitable. Bodmer (Bodmer, 1965; Bodmer and Laird, 1968) investigated the effect of DNA synthesis on integration of donor DNA during transformation in *B. subtilis*. For this purpose competent cells of the thy^- strain, unable to synthesize thymine, were transferred at the moment of addition of DNA to them to a medium containing 5BU. The density of the donor and recipient DNA isolated after growth for different periods on this medium was determined by centrifugation in a density gradient. The transforming activity of DNA isolated from the cells after incubation for 30 min with the donor DNA was found to remain predominantly in the light peak, although about 10% of it is transferred to the intermediate fraction between the light and hybrid DNAs. This transfer becomes particularly conspicuous after further incubation for 45 min in the medium with 5BU. If in that case the ordinary semiconservative synthesis of the whole chromosome had taken place, donor activity would have been observed in the hybrid fraction. In fact, however, the DNA molecules have intermediate density, as though only part of one of the complementary strands had been synthesized afresh in the medium with 5BU. Since the molecular weight of DNA isolated from the cells of *B. subtilis* varies between 6×10^6 and 30×10^6 daltons, the extent of reparative synthesis can be estimated to be from 2000 to 10,000 nucleotides, in good agreement with the value calculated on the basis of the genetic data.

Recombination at the half-chiasma level during transformation automatically leads to integration of the exogenote (i.e., double exchange). Odd exchanges simply cannot be explained by this mechanism. What happens during conjugation? If recombination in this case also takes place at the single-strand level (Kunicki-Goldfinger, 1968; Curtiss, 1969) the problem of accounting for double exchanges is removed. However, this mechanism appears unlikely, first of all because the physical scale of the map during conjugation is an order of magnitude higher than during transformation.*

---

*Evidence that the exogenote becomes double-stranded (see page 148) before recombination is given by the results of experiments by Bresler et al. (1971) which showed that rec^- merozygotes can carry out a low-grade induced synthesis of β-galactosidase independently of Hfr selection (i.e., with transfer of both left and right strands).

Admittedly a sharp increase in the frequency of genetic exchanges is observed at the ends of the exogenote. It may be that recombination in this case actually takes place principally by semicrossing-over, resulting in the insertion of small single-stranded segments of donor DNA into the chromosome of the recipient cell. If a site of heterozygosity is formed, a correction mechanism is activated, leading to nonreciprocity (Herman, 1968a,b). Evidence that the processes taking place during chiasma formation are universal is also given by the high negative interference observed in *E. coli* (Jacob and Wollman, 1961: 295; Norkin, 1970) and the recently discovered polarity of intra-operon recombination in *Streptomyces coelicolor* (Sermonti and Carere, 1968).

According to a preliminary communication (Matney et al., 1971) recombination in a partially diploid strain of *E. coli* takes place on account of semicrossing-over induced in the operator of the *his* operon and accompanied by correction. The hybrid DNA under these circumstances may occupy the whole operon.

Whereas recombination during conjugation takes place at the double-helical level, double crossing-over is evidently necessary for integration of the exogenote. An odd number of exchanges would lead to the formation of an open chromosome with terminal redundancies (Hopwood, 1967). However, this statement is valid only for open structures. In the genetics of microorganisms systems are known in which integration takes place as the result of one crossing-over, since the structures which recombine are not linear (open) but circular. Examples are the integration (lysogenization) of temperate phages in the cell chromosome (Campbell, 1962; Signer and Rybchin, 1967) and interaction between F factor and the chromosome (Scaife, 1966; Sharp et al., 1972). By analogy with this it can be postulated (Bresler et al., 1967) that during conjugation the DNA fragments in the bacterial cell may form a closed ring (Leavitt et al., 1971). An even number of exchanges between the ring and the homologous region of the bacterial chromosome leads to the formation of a recombinant chromosome and recombinant ring. The latter can be eliminated soon after. An odd number of exchanges leads to elimination of a fragment into the chromosome and thus gives rise to redundancy. This type of structure must evidently be unstable and form loops within which further exchanges are possible, leading ultimately to a stable state. This mechanism also explains the heterogeneity of the progeny of individual conjugating pairs. Heterogeneity of the progeny of exconjugants means that each zygote can give more than one recombinant type, and elimination of repeated recombinant types is observed in different generations (from the 3rd to the 9th) of the same line (Anderson, 1958). The results obtained by statistical analysis of the progeny of exconjugants in a system of closely linked markers are in agreement with the model of successive incorporation (Bresler et al., 1970, 1973).

On the other hand, results obtained by selection within a narrow interval indicate that the exogenote can reproduce as a linear fragment (Lotan et al., 1972; Frédéricq, 1972). If this is so, the existence of another mechanism responsible for the evenness of the exchanges will have to be postulated. This mechanism could be an increase in recombination at the ends of the exogenote through the action of the *recBC* system.

High Negative Interference in Three-Point Crosses. Two types of experimental results have been described in terms of interference: the character of distribution of the flanks in a standard flank system (2.8) and the formation of wild-type recombinants in three-point crosses (2.9). The simplest and most natural explanation of the character of distribution of the flanks is given by Holliday's hypothesis. Can this hypothesis explain the phenomenon of high negative interference observed in three-point crosses? To answer this question, we must analyze how wild-type recombinants can arise in three-point crosses. They evidently can do so either as the result of a double half-chiasma (double semicrossing-over, Fig. 94) or as the result of absence of linked correction during the formation of a half-chiasma between the terminal and central sites (Fig. 95). The probability of double and semicrossing-over evidently cannot be greater than the product of the probabilities of single semicrossovers. There remains the second mechanism. In two-point crosses rf is a component of two probabilities: the probability that only one of the markers (p_a) is incorporated into the hybrid DNA and the probability of unlinked correction (p_u) if both markers were incorporated into the hybrid region (p_{ab}). Since rf is determined not only by physical distance, but also by the degree of directivity of correction (d), the equation for rf assumes the following form:

$$rf_{ab} = p_a d^{-1} + p_{ab} p_u d^{-1}.$$

Let us assume for simplicity that $rf_{ab} = rf_{bc}$ and $p_a = p_{ab}$ (as is the case, for example, with alleles *1* and *2* of the arg_4 gene, Table 23). The equation then becomes

$$rf_{ab} = p_a d^{-1}(1 + p_u).$$

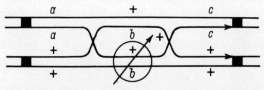

Fig. 94. Possible mechanism of formation of wild-type recombinants in three-point ($ac \times b$) crosses. Legend as in Figs. 72 and 74.

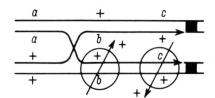

Fig. 95. Mechanism of high negative interference. Correction of the $c/+$ site is not essential. Legend as in Figs. 72 and 74.

The "theoretically" expected probability of appearance of wild-type recombinants ($rf_{ab} \times rf_{bc}$) is $p_a^2 d^{-2}(1+p_u)^2$, while according to the proposed scheme it is $p_a p_u d^{-1}$. Consequently

$$C = \frac{p_a p_u d^{-1}}{p_a^2 d^{-2}(1+p_u)^2} = \frac{p_u}{p_a d^{-1}(1+p_u)^2}.$$

Since all the parameters are much less than unity (if directed correction takes place), $C > 1$. For example, if $p_u = p_a = d^{-1} = 0.1$, $C = 10$.

It follows from this equation that high negative interference must be allele-specific (for example, it is not observed in crosses involving markers which reduce the degree of directivity of correction to zero). In fact, it has been clearly shown in at least one case that the value of C is reduced by several times if, instead of a point mutation, the central component in the three-point cross is a deletion (Berger and Warren, 1969; see also Goldman and Gutz, 1971). Further support for this hypothesis is also given by the wild scatter of values of C in different crosses (Meselson, 1965; Doermann and Parma, 1967; Matvienko, 1968, 1969). It also follows from the model that wild-type recombinants in three-point crosses must appear predominantly with the P configuration of the flank markers. The proposed mechanism of formation of double recombinants in three-point crosses is confirmed by results (Kruszewska and Gajewski, 1967) showing that octads of the 6– : 2+ type appear in such crosses as a result of conversion of the central allele to the wild type.

The absence of interference during transformation (Gray and Ephrussi-Taylor, 1967) shows that in this case double recombinants are formed as a result of double semicrossing-over. Consequently, the breakage sites are not fixed on the chromosome.

Correction and Repair. The idea of correction arose in connection with the study of the nature of the appearance of pure clones of mutants in experiments on induced mutagenesis. It has been known for a long time that the most widely different mutagens induce both complete and partial mutations. In the first case the mutable unit is the whole chromatid, while in the second case it is the half-chromatid, with the result that mosaic mutants are formed.

Before the structure of DNA had been established attempts were made to find an explanation for the segregation of mutations, but when proof of the semiconservative character of DNA replication was obtained the problem was transferred to a different plane. Clearly in most cases the mutation affects only one of the complementary DNA strands and, consequently, a molecular heterozygote is formed. After semiconservative replication such a molecule must "segregate" in the genetic sense of the term, i.e., it must give rise to two populations of molecules with different genetic properties (a mixed clone). How, in this case, can pure clones appear? Several hypotheses have been proposed (survey: Nasim and Auerbach, 1967):

1. The companion lethal hypothesis, the essence of which is that the second strand suffers a lethal injury and does not replicate.
2. The simultaneous hit hypothesis: the injury leading to mutation affects both complementary DNA strands.
3. The "master strand" hypothesis: genetic information is transmitted by only one of the DNA strands (the master strand); if a mutation arises in it, it is transmitted in the first replication cycle to both daughter molecules.
4. The "incidental" repair hypothesis: homozygosis of mutation damage takes place incidentally with repair of lethal injuries (by analogy with the mechanism of linked correction).
5. The correction hypothesis: heterozygosis is abolished by elimination of one of the unpaired bases and subsequent reparative replication. The correction mechanism is at least partly responsible for homozygosis of mutations induced by UV irradiation (Haefner, 1967).

At this point differences in the genetic meaning of the mechanisms of correction and repair must be emphasized.* Repair always implies removal of an injury and restoration of the original state (Carrier and Setlow, 1970; Paribok and Tomilin, 1970). Correction can take place in any direction.

The remarkable analogy between the repair and correction mechanisms suggests that they may be controlled by the same enzyme system. However, in transformation (Bresler et al., 1968) and mutation (Kubitschek, 1966) heterozygotes no effect of uvr^- mutations on homozygosis was found (this, by the way, is evidence against the hypothesis of incidental repair). Further, in the xrs_1 strain of yeast with an unidentified dark repair defect the fraction of complete mutations is twice as large as in the wild type (Arman et al., 1972). The search for strains of fungi, sensitive to UV irradiation, in which spontaneous conversion is suppressed likewise has not yet proved successful

---

*The term "repair" is used as an abbreviation for "repair by excision and substitution" (or "repair by substitution and excision").

(survey: Holliday, 1968). On the other hand, *hex* mutants (page 215) have been found to be resistant to irradiation. Correctase and UV-specific endonuclease are unquestionably different enzymes. Holliday and Hollaman (1972) characterized the deoxyribonuclease absent in the nuc_2^- strain of Ustilago, in which conversion is suppressed but crossing over takes place normally (Badman, 1972). At the same time, the evidence obtained by the study of transformation and phage heterozygotes indicates that there is some sort of link between correction and repair. If donor DNA is irradiated before transformation with large doses of UV light all the transformants give mixed clones (Bresler et al., 1971). In a study of heterozygotes of phage λ, Kellenberger et al. (1962) found that UV irradiation of bacteria infected with crossed phages increases the frequency of formation of heterozygous particles by more than one order of magnitude. With optimal doses (at which the yield of phages per infected bacterial cell does not exceed 1-3 particles) the frequency of heterozygotes reaches 1%. This fact, which was confirmed by other experiments with UV (Doerfler and Hogness, 1968) and mitomycin (Baas and Jansz, 1972), is interpreted as the binding of certain enzymes by the damaged chromosome which limit the correction process and it suggests that the mechanisms of correction and repair overlap to some extent.

TABLE 33. Phenotypic Effect of the Blocking of Different Stages of Recombination by Mutation (after Holliday, 1968)

Stages	Phenotypic effect		
	Crossing-over	Conversion	Repair
1. Regulation of stages 2-5	$+^a$	$+^a$	+
2. Pairing	−	−	+
3. Endonucleotic breakage in the recombiner	$-^b$	$-^b$	+
4. Dissociation	−	−	+
5. Endonucleotic breakage of the half-chiasma	−	$+^c$	+
6. Reunion of phosphate bonds by ligase	−	−	−
7. Correction	+	$-^d$	+
8. Exonucleotic widening of the gaps	−	−	−
9. Reparative synthesis	−	−	+
Excision of dimers	+	+	−

[a] Constitutive mutants with a high frequency of mitotic recombination can be found if a low frequency is connected with repression of the corresponding enzymes in mitosis.
[b] Mutations of the recombiner must affect recombination in one region only.
[c] In the absence of secondary breakages the temporary formation of hybrid DNA favors conversion.
[d] If the correction mechanism is completely blocked, the frequency of postmeiotic segregation must be sharply increased.

Holliday (1968) has drawn up a table of possible correlations between the *uvr⁻* phenotype and recombinations. I have modified it to make it correspond to my own hypothesis (Table 33, 7-9).

5.7. Conclusion

During the last 10-15 years striking progress has been made toward a solution of the recombination problem, and a real breakthrough has been made literally in the last 3-5 years. As long ago as in 1964 Hayes wrote that "... its nature (the nature of recombination) remains one of the few dark corners of cellular behavior which the recent developments in molecular biology have, so far, done little to illuminate. In a sense, our concepts about the mechanism of recombination are more confused now than they were a quarter of a century ago ..." (Hayes, 1968, p. 376). For a breakthrough in this field a surge of thinking on the theoretical plane and the development of a new system of ideas were essential: the experimental facts were already more than sufficient even at that time. That this conclusion is correct is confirmed by the fact that when the words cited above were being written, Holliday and Whitehouse had formulated a number of new theoretical concepts which provided a basis for further progress.

It must not be forgotten, however, that theoretical advances in any field of science (except mathematics) are impossible without fundamental experimental research. Theoretical concepts could not be successfully put forward until (1) genetic analysis had provided information on the phenomenological aspects of recombination between within genes, (2) proof of the chiasmatype theory had been obtained by cytogenetic methods, (3) the structure of DNA and of chromosomes had been decoded by physicochemical and biochemical methods, (4) the structure of recombinant DNA molecules had been studied by the same methods, and (5) the enzymic character of recombination had been established by radiation-genetic and biochemical methods and the enzymes participating in it had been identified.

We now know that recombination is an enzymically controlled process taking place at the level of physical interaction between DNA molecules. An essential (but not sufficient in itself) condition for this interaction is union of the homologous chromosomes in a single cell (complete or partial zygote). Another condition is their specific pairing both at the supermolecular (synaptinemal complex) and at the molecular level (DNA-binding protein). Recombination takes place by physical breakages and reunions of DNA molecules. Breakages and reunions may occur in both strands of the double-helical molecule (crossing-over) or in only one strand (semicrossing-over). Both crossing-over and semicrossing-over are accompanied by the formation of hybrid DNA regions at sites of interaction between chromosomes. In the absence of correction of molecular heterozygosity this would lead to frequent cases of mosaicism (postmeiotic segregation). The mechanism of cor-

CONCLUSION

rection removes this threat and at the same time leads to nonreciprocity of intragenic recombinations (conversions). A differential action on intragenic and intergenic recombination is possible because of the existence of this mechanism. It determines the allele specificity of intragenic recombination and some of its other special features.

In the most primitive systems (transformation of phages by single-stranded DNA, transformation of bacteria and, perhaps, certain types of transduction) recombination, both intragenic and intergenic, takes place as the result of semicrossing-over. Moreover, the endonuclease which generates the primary breakages evidently does not possess site specificity (there are no fixed breakage points). For these reasons in such systems neither polarity nor interference is observed. The presence of the correction mechanism is manifested by the allele specificity of recombination.

Both intergenic and intragenic recombination in phages take place by two mechanisms: by reciprocal crossing-over and by semicrossing-over. The presence of semicrossing-over and correction are revealed by an increase in the frequency of exchanges at the ends of the chromosomes, high negative interference, and allele specificity of recombination at short distances, and nonreciprocity even if allowance is made for intergenic recombination.

In the temporate phages an additional mechanism of recombination appears. This is site-specific crossing-over, by means of which the phage chromosome is integrated with the bacterial genome. The enzyme controlling the Int system possibly possesses two binding centers and endonucleotic activity arises only after its attachment to the corresponding specific sites on both recombining chromosomes. These sites can be regarded as analogues of the recombiners in fungi. Information on genetic effects (such as interference, polarity, and nonreciprocity) for conjugation systems is not yet sufficiently reliable. Nevertheless, two mechanisms of recombination — reciprocal crossing-over and semicrossing-over — can be considered to operate in them.

In eukaryotes intergenic recombination takes place through controlled site-specific crossing-over. Intragenic recombination takes place by both these mechanisms.

Although the details of several recombination models have now been worked out and further research is required before the final choice between them can be made, it seems certain that the general picture drawn above is correct and that subsequent development will take place within the framework of this system of ideas.

The main task facing geneticists at present is the study of the regulation of recombination processes in the genome and of the behavior of the chromosome as a whole in recombination. I am confident that research along these lines will elucidate the molecular mechanisms of some of the problems in classical genetics such as the Schultz-Redfield effect, synapsis, and chromosome interference.

Bibliography

Abel, W.O., 1965, Z. Vererbungsl., 96:228.
Abel, W.O., 1967, Ber. Dtsch. bot. Ges., 80:128.
Abel, W.O., 1971, Mol. Gen. Genet., 110:61.
Adams, A., 1972, Mol. Gen. Genet., 118:311.
Adler, K., K. Beyreuther, E. Fanning, N. Geisler, B. Gronenborn, A. Klemm, B. Muller-Hill, M. Pfahl, and A. Schmitz, 1972, Nature, 237:322.
Agol, I.I., 1929, Zh. Eksperim. Biol. Ser. A, 5:86.
Ahmad, A.F., D. J. Bond, and H.L.K. Whitehouse, 1972, Genet. Res., 19:121.
Alberts, B.M., 1970, Federat. Proc., 29:1154.
Alberts, B.M., and L. Frey, 1970, Nature, 227:1313.
Alberts, B., L. Frey, and H. Delius, 1972, J. Mol. Biol., 68:138.
Amati, P.A., and M. Meselson, 1965, Genetics, 51:369.
Anagnostopoulos, C., and J. P. Crawford, 1961, Proc. Nat. Acad. Sci. USA., 47:1900.
Anderson, E.G., 1925, Genetics, 10:403.
Anderson, E.G., 1936, in: Biological Effects of Radiation, New York, p. 1297.
Anderson, T.F., 1958, Cold Spring Harbor Sympos. Quant. Biol., 23:47.
Angel, T., B. Austin, and D.G. Catcheside, 1970, Austral. J. Biol. Sci., 23:1229.
Anraku, N., and I.R. Lehman, 1969, J. Mol. Biol., 46:467.
Anraku, N., and J. Tomizawa, 1964a, J. Mol. Biol., 8:508.
Anraku, N., and J. Tomizawa, 1964b, J. Mol. Biol., 8:516.
Anraku, N., and J. Tomizawa, 1965, J. Mol. Biol., 11:501.
Anraku, N., Y. Anraku, and I.R. Lehman, 1969, J. Mol. Biol., 46:481.
Ansley, H.R., 1958, J. Biophys. Biochem. Cytol., 4:59.
Arber, W., 1964, Annual Rev. Microbiol., 19:365.
Arber, W., 1971, in: The Bacteriophage Lambda, Cold Spring Harbor Lab., p. 90.
Arman, I.P., A.B. Devin, and T.A. Dutova, 1972, Genetika, 8(12):174.
Arwert, F., and G. Venema, 1973, in: Proc. 1st Europ. Congress on Transformation, Academic Press, New York.
Baas, P.D., and H.S. Jansz, 1972, J. Mol. Biol., 63:557.
Badley, M.W., 1968, Cold Spring Harbor Sympos. Quant. Biol., 33:333.
Badman, R., 1972, Genet. Res., 20:213.
Bailey, N.T.I., 1951, Heredity, 5:111.
Bailey, N.T.I., 1961, Introduction to the Mathematical Theory of Genetic Linkage, Oxford, 119:136.
Baker, B.S., and A.T. Carpenter, 1972, Genetics 71:255.
Baker, R.M., and R.H. Haynes, 1972, Virology, 50:11.
Baker, W.K., and J.A. Swatek, 1965, Genetics, 52:191.

Balbinder, E., 1962, Genetics, 47:545.
Baranowska, H., 1970, Genet. Res., 16:185.
Barbour, S.D., 1972, Mol. Gen. Genet., 117:303.
Barbour, S.D., and A.J. Clark, 1970, Proc. Nat. Acad. Sci. USA. 65:955.
Barbour, S.D., H. Nagaishi, A. Templin, and A.J. Clark, 1970, Proc. Nat. Acad. Sci. USA, 67:128.
Barry, E.G., 1966, Genetics, 54:321.
Barry, E.G., 1969, Chromosoma, 26:119.
Bateson, W., E.R. Saunders, and R.C. Punnett, 1905, Rept. Evol. Comm. Roy. Soc., 2:1.
Bausum, H.T., and R.P. Wagner, 1965, Genetics, 51:815.
Bautz, E.K.F., F.A. Bautź, and W. Ruger, 1968, Cold Spring Harbor Sympos. Quant. Biol., 33:59.
Bautz, F.A., and E.K.F. Bautz, 1967, Genetics, 57:887.
Bayer, A., 1965, Chromosoma, 17:291.
Bayliss, M.B., and R. Riley, 1972, Genet. Res., 20:193.
Bayliss, M.B., and R. Riley, 1973, Heredity, 30:258.
Beadle, G.W., 1930, Cornell. Univ. Agr. Exper. Stat. Mem., 129:1.
Beadle, G.W., 1932, Genetics, 17:481.
Beadle, G.W., 1933, Cytologia, 4:269.
Beadle, G.W., 1957, in: The Chemical Basis of Heredity, Baltimore, p.2.
Beadle, G.W., and S. Emerson, 1935, Genetics, 20:192.
Beattie, K.L., and J.K. Setlow, 1971, Nature New Biol., 231:177.
Beccari, E., P. Modigliani, and G. Morpurgo, 1967, Genetics, 56:7.
Beckendorf, S.K., and J.H. Wilson, 1972, Virology, 50:315.
Belling, J., 1927, J. Genetics, 18:177.
Belling, J., 1928, Univ. Calif. Publs Bot., 14:283.
Belling, J., 1933, Genetics, 18:388.
Benz, W.C., and H. Berger, 1973, Genetics, 73:1.
Benzer, S., 1955, Proc. Nat. Acad. Sci. USA, 41:344.
Benzer, S., 1957, in: The Chemical Basis of Heredity, Baltimore, p. 70.
Berg, C.M., and R. Curtiss, 1967, Genetics, 56:503.
Berg, D.E., 1971, in: The Bacteriophage Lambda, Cold Spring Harbor Lab., p. 667.
Berg, D.E., and J.A. Gallant, 1971, Genetics, 68:457.
Berger, H., 1965, Genetics, 52:729.
Berger, H., and A.J. Warren, 1969, Genetics, 63:1.
Berger, H., and A.W. Kozinski, 1969, Proc. Nat. Acad. Sci. USA, 64:897.
Berger, H., A.J. Warren, and K.E. Fry, 1969, J. Virology, 3:171.
Bernstein, H., 1962, J. Theoret. Biol., 3:335.
Bernstein, H., 1964, J. Theoret. Biol., 6:347.
Bernstein, H., 1967, Genetics, 56:755.
Bernstein, H., 1968, Cold Spring Harbor Sympos. Quant. Biol., 33:325.
Bird, M.J., and D.G. Fahmy, 1953, Proc. Roy. Soc. B, 140:556.
Bloch, D.P., and G.C. Godman, 1955, J. Biophys. Biochem. Cytol., 1:17.
Bodmer, W.F., 1965, J. Mol. Biol., 14:534.
Bodmer, W.F., and A.I. Darlington, 1969, in: Genetic Organization, 1, New York–London, p. 223.
Bodmer, W.F., and A.T. Ganesan, 1964, Genetics, 50:717.
Bodmer, W.F., and C.D. Laird, 1968, in: Replication and Recombination of Genetic Material, Canberra, p. 184.
Bogdanov, Yu. F., A.B. Iordanskii, and V. M. Gindilis, 1965, Genetika, 5:82.
Bole-Gowda, B.N., D.D. Perkins, and W.N. Strickland, 1962, Genetics, 47:1243.
Bonner, D.M., 1951, Cold Spring Harbor Sympos. Quant. Biol., 16:143.

Boon, T., and N.D. Zinder, 1969, Proc. Nat. Acad. Sci. USA, 64:573.
Boon, T., and N.D. Zinder, 1971, J. Mol. Biol., 58:133.
Botstein, D., and M. Matz, 1970, J. Mol. Biol., 54:417.
Boveri, Th., 1902, Verhandl. Phys.-Med. Ges. Würzburg, N.F., 35:67.
Boyce, R.P., and P. Howard-Flanders, 1964, Proc. Nat. Acad. Sci. USA, 51:293.
Brandham, P.E., 1969, Chromosoma, 26:270.
Brenner, S., and C. Mulstein, 1966, Nature, 211:242.
Brent, T.P., 1972, Nature, 239:172.
Bresch, C., 1955, Z. Naturforsch., 106:545.
Bresler, S.E., 1966, Introduction to Molecular Biology [in Russian], 2nd Edition, Moscow–Leningrad.
Bresler, S.E., and V.A. Lantsov, 1966, Genetika, 8:83.
Bresler, S.E., R.A. Kreneva, V.V. Kushev, and M.I. Mosevitskii, 1964a, Biokhimiya, 29:477.
Bresler, S.E., R.A. Kreneva, V.V. Kushev, and M.I. Mosevitskii, 1964b, Biokhimiya, 29:1103.
Bresler, S.E., R.A. Kreneva, V.V. Kushev, and M.I. Mosevitskii, 1964c, Z. Vererbungsl., 95:288.
Bresler, S.E., R.A. Kreneva, V.V. Kushev, and M.I. Mosevitskii, 1964d, J. Mol. Biol., 8:79.
Bresler, S.E., V.L. Kalinin, and D.A. Perumov, 1966, Genetika, 5:133.
Bresler, S.E., V.A. Lantsov, and A.A. Blinkova, 1967, Genetika, 8:73.
Bresler, S.E., R.A. Kreneva, and V.V. Kushev, 1968, Mol. Gen. Genetics, 102:257.
Bresler, S.E., L.P. Dadivanjan, and M.I. Mosevitsky (Mosevitskii), 1970, Biochim. Biophys. Acta, 224:249.
Bresler, S.E., V.A. Lantsov, and L.R. Manukyan, 1970, Genetika, 8:116.
Bresler, S.E., R.A. Kreneva, and V.V. Kushev, 1971, Mol. Gen. Genetics, 113:257.
Bresler, S.E., V.A. Lantsov, and A.A. Luk'yanets-Blinkova, 1971, Molekul. Biol. 5:803.
Bresler, S.E., L.G. Vyacheslavov, V.L. Kalinin, R.A. Kreneva, V.V. Kushev, V.A. Lantsov, M.I. Mosevitskii, D.A. Perumov, and V.N. Rybchin, 1973, Elementary Processes of Genetics (In Russian), Leningrad, Nauka, p. 116.
Brewen, J.G., and W.J. Peacock, 1969, Proc. Nat. Acad. Sci. USA, 62:389.
Bridges, B.A., 1971, J. Bacteriol., 108:944.
Bridges, B.A., and R.J. Munson, 1966, Biochem. Biophys. Res. Commun., 22:268.
Bridges, C.B., 1915, J. Exper. Zool., 29:1.
Bridges, C.B., 1916, Genetics, 1:107.
Bridges, C.B., 1917, Amer. Naturalist, 51:370.
Bridges, C.B., and E.G. Anderson, 1925, Genetics, 10:418.
Broda, P., J.R Beckwith, and J. Scaife, 1964, Genet. Res., 5:489.
Broker, T.R., and I.R. Lehman, 1970, Genetics, 64, 2, Part 2:58.
Broker, T.R., and J.R. Lehman, 1971, J. Mol. Biol., 60:131.
Brown, S.W., and D. Zohary, 1955, Genetics, 40:850.
Browning, L.S., and E. Altenburg, 1964, Genetics, 50:695.
Brunk, C.F., and P.C. Hanawalt, 1967, Science, 158:663.
Brunswik, H., 1926, Z. Bot., 18:481.
Burnham, C.R., J.T. Stout, W.H. Weinheimer, R.V. Kowles, and R.L. Phillips, 1972, Genetics, 71:111.
Buttin, G., and M. Wright, 1968, Cold Spring Harbor, Symp. Quant. Biol., 33:259.
Cairns, J., 1963, J. Mol. Biol., 6:208.
Calef, E., 1957, Heredity, 11:265.
Callan, H.G., 1960, Internat. Rev. Cytol., 16:1.
Callan, H.G., 1967, J. Cell Sci., 2:1.

Callan, H.G., and L. Lloyd, 1956, Nature, 178:355.
Callan, H.G., and L. Lloyd, 1960, Philos. Trans. Roy. Soc. London B., 24:135.
Cameron, H.R., K.S. Hsu, and D.D. Perkins, 1966, Genetics, 57:1.
Campbell, A., 1962, Advances in Genetics, 11:101.
Campbell, A., 1963, Virology, 20:344.
Campbell, D.A., 1968, Genetics, 60:166.
Campbell, J.L., L. Soll, and C.C. Richardson, 1972, Proc. Nat. Acad. Sci. USA, 69:2090.
Carlson, E., 1959, Quart. Rev. Biol., 34:33.
Carlson, K., and A.W. Kozinski, 1970, J. Virology, 6:344.
Carlson, P.S., 1971, Genet. Res., 17:53.
Carlson, P.S., 1972, Genet. Res., 19:129.
Carlton, B.C., 1966, J. Bacteriol., 91:1795.
Carrier, W.L., and R.B. Setlow, 1970, J. Bacteriol., 102:178.
Carrol, R.B., K.E. Neet, and D.A. Goldthwait, 1972, Proc. Nat. Acad. Sci. USA, 69:2741.
Carter, D.M., and C.M. Radding, 1971, J. Biol. Chem., 246:2502.
Case, M.E., and N.H. Giles, 1958, Cold Spring Harbor Sympos. Quant. Biol., 23:119.
Case, M.E., and N.H. Giles, 1964, Genetics, 49:529.
Cassuto, E., and C.M. Radding, 1971a, Nature New Biol., 229:13.
Cassuto, E., and C.M. Radding, 1971b, Nature New Biol., 230:128.
Cassuto, E., T. Lash, K.S. Sriprakash, and C.M. Radding, 1971, Proc. Nat. Acad. Sci. USA, 68:1639.
Castellazzi, M., J. George, and G. Buttin, 1972, Mol. Gen. Genet., 119:153.
Caster, J.H., and S.H. Goodgal, 1972, J. Bacteriol., 112:492.
Castle, W.F., 1919, Proc. Nat. Acad. Sci. USA, 5:25.
Catcheside, D.G., 1966a, Austral. J. Biol. Sci., 19:1039.
Catcheside, D.G., 1966b, Austral. J. Biol. Sci., 19:1047.
Catcheside, D.G., 1968, Genetics, 59:443.
Catcheside, D.G., and B. Austin, 1969, Amer. J. Bot., 56:685.
Catcheside, D.G., and B. Austin, 1971, Austral. J. Biol. Sci., 24:107.
Catcheside, D.G., A.P. Jessop, and B.R. Smith, 1964, Nature, 202:1242.
Cavalieri, L.F., and B.H. Rosenberg, 1961, Biophys. J., 1:337.
Cato, A., and W.R. Guild., 1968, J. Mol. Biol., 37:157.
Celis, J.E., J.D. Smith, and S. Brenner, 1973, Nature New Biol., 241:130.
Chang, L.-T., and R.W. Tuveson, 1967, Genetics, 56:801.
Chase, M., and A.H. Doermann, 1958, Virology, 43:332.
Chernik, T.P., and A.S. Kriviskii, 1968, Genetika, 6:75.
Chiang, K.S., J.R. Kates, and N. Sueoka, 1965, Genetics, 52:442.
Childs, J.D., 1971, Genetics, 67:455.
Chinnici, J.P., 1971, Genetics, 69:85.
Chiu, S.M., and P.J. Hastings, 1973, Genetics, 73:29.
Chovnick, A., C.H. Ballantyne, D.L. Baillie, and D. G. Holm, 1970, Genetics, 66:315.
Chovnick, A., G.H. Ballantyne, and D.G. Holm, 1971, Genetics, 69:179.
Clark, A.J. 1971, Annual Rev. Microbiol. 25:437.
Clark, A.J., 1972, Lunteren Lectures on Molecular Genetics, Vol. 1.
Clark, A.J., and A.D. Margulies, 1965, Proc. Nat. Acad. Sci. USA, 53:451.
Clarkson, J.M., and H.J. Evans, 1972, Mut. Res., 14:413.
Clavilier, L., M. Luzzatti, and P.P. Slonimski, 1960, Compt. Rend. Soc. Biol., 154:1970.
Cleaver, J.E., 1968, Nature, 218:652.
Cleaver, J.E., 1971, Mut. Res., 12:453.
Cole, A., 1962, Nature, 196:211.
Cole, R.S., 1971, J. Bacteriol., 106:143.
Cole, R.S., 1973, Proc. Nat. Acad. Sci. USA, 70:1064.

Coleman, J.R., and M.J. Moses, 1964, J. Cell Biol., 23:63.
Collins, C.J., and W.R. Guild, 1972, J. Bacteriol., 109:266.
Comings, D.E., 1971, Nature New Biol., 229:24.
Comings, D.E. 1972, Expt. Cell Res., 74:383.
Comings, D.E., and A.G. Motulsky, 1966, Blood, 28:54.
Comings, D.E., and T.A. Okada, 1969, Biophys, J., 9, Abstr., A19.
Comings, D.E., and T.A. Okada, 1970, Nature, 227:451.
Cooper, G.M., and M.S. Fox, 1969, Biochem. Biophys. Res. Commun., 34:777.
Cooper, K.W., 1949, J. Morphol., 84:81.
Cooper, P.K., and P.C. Hanawalt, 1972a, J. Mol. Biol., 67:1.
Cooper, P.K., and P.C. Hanawalt, 1972b, Proc. Nat. Acad. Sci. USA, 69:1156.
Correns, C., 1909, Z. Vererbungsl., 1:291.
Couch, J., and P.C. Hanawalt, 1967, Biochem. Biophys. Res. Commun., 29:779.
Coukell, M.B., and C. Yanofsky, 1971, J. Bacteriol., 105:864.
Couzin, D.A., and D.P. Fox, 1973, Chromosoma, 41:421.
Crawford, I.P., and J. Preiss, 1972, J. Mol. Biol., 71:717.
Creighton, H.B., and B. McClintock, 1931, Proc. Nat. Acad. Sci. USA, 17:492.
Csordas, A., D.M. Mount, S.D. Barbour, and D.A. Goldthwait, 1972, Lunteren Lectures on Molecular Genetics, S2:3.
Cuénot, L., 1904, Arch. Zool. Exptl. et Gén. Notes et Revue, 2:45.
Curtiss, R., 1968, Genetics, 58:9.
Curtiss, R., 1969, Annual Rev. Microbiol., 23:69.
Dalrymple, G.V., J.L. Sanders, and M.L. Baker, 1968, J. Theoret. Biol., 21:368.
Darlington, C.D., 1934a, Genetics, 19:95.
Darlington, C.D., 1934b, Z. Vererbungsl., 67:96.
Darlington, C.D., 1937, Recent Advances in Cytology. 2nd ed. London.
Darlington, C.D., 1940, J. Genetics, 39:351.
Davidoff-Abelson, R., and D. Dubnau, 1971, Proc. Nat. Acad. Sci. USA, 68:1070.
Davies, D.R., 1966, Z. Vererbungsl., 98:61.
Davies, D.R., and H.J. Evans, 1966, Advances in Radiat. Biol., 2:243.
Davies, D.R., and C.W. Lawrence, 1967, Mutat, Res., 4:147.
Davies, R.W., and J.S. Parkinson, 1971, J. Mol. Biol., 56:403.
Davis, R.W., 1968, Biophys. J., 8, Abstr., TA 13.
Davis, X.J., 1972, Lunteren Lectures on Molecular Genetics, S3:3.
De, D.N., 1968, 12th Internat. Congr. Genet., Part II, Tokyo, p. 72.
Delbrück, M., and G. Stent, 1957, in: The Chemical Basis of Heredity, Baltimore, p. 562.
Delius, H., N.J. Mantell, and B. Alberts, 1972, J. Mol. Biol., 67:341.
Demerec, M., 1928, Genetics, 13:359.
Demerec, M., 1962, Proc. Nat. Acad. Sci. USA, 48:1696.
Demerec, M., J. Goldman, and E.L. Lahr, 1958, Cold Spring Harbor Sympos. Quart. Biol., 23:59.
Detlefsen, J.A., and E.J. Roberts, 1921, J. Exper. Zool., 32:333.
De Vries, H., 1903, Befruchtung und Bastardierung, Leipzig.
DeVries, J.K., and W.K. Maas, 1971, J. Bacteriol. 106:150.
Dickinson, S., 1928, Proc. Roy. Soc. B, 103:547.
Digby, L., 1910, Ann. Bot., 24:727.
Dobzhansky, T., 1929, Biol. Zbl., 49:408.
Doerfler, W., and D.S. Hogness, 1968, J. Mol. Biol., 33:661.
Doermann, A.H., and L. Boehner, 1963, Virology, 21:551.
Doermann, A.H., and M. B. Hill, 1953, Genetics, 38:79.
Doerman, A.H., and D.H. Parma, 1967, J. Cell. Comp. Physiol., 70, suppl. 1, p. 147.
Dorp, B. van, 1972, Lunteren Lectures on Molecular Biology, SXI:1.

Doty, P., J. Marmur, J. Eigner, and C. Schildkraut, 1960, Proc. Nat. Acad. Sci. USA., 46:461.
Douglas, L.T., and H.W. Kroes, 1969, Genetica, 40:503.
Dover, G.A., and R. Riley, 1972, Nature New Biol., 235:61.
Drake, J. W., 1966, Proc. Nat. Acad. Sci. USA, 55:506.
Drake, J.W., 1967, Proc. Nat. Acad. Sci. USA, 58:962.
Drapeau, G.R., W.J. Brammar, and C. Yanofsky, 1968, J. Mol. Biol., 35:357.
Dubinin, N.P., 1929, Zh. Éksperim. Biol., ser A, 5:53.
Dubinin, N.P., N. Sokolov, and G. Tiniakov, 1937, Drosophila Inform. Service, 8:76.
Dubinin, N.P., L.S. Nemtseva, and V.P. Romanov, 1972, Genetika, 8(12):98.
Dubnau, D., and C. Cirigliano, 1972a, J. Mol. Biol., 64:9, 31.
Dubnau, D., and C. Cirigliano, 1972b, J. Bacteriol., 111:488.
Dubnau, D., and C. Cirigliano, 1973a, Molec. Gen. Genet., 120:101.
Dubnau, D., and C. Cirigliano, 1973b, J. Bacteriol. 113:1512.
Dubnau, D., and R. Davidoff-Abelson, 1971, J. Mol. Biol., 56:209.
Dubnau, D., R. Davidoff-Abelson, and I. Smith, 1969, J. Mol. Biol., 45:155.
Dulbecco, R., 1964, J. Cellular and Compar. Physiol., 64, suppl. 1, p. 181.
Du Praw, E.J., 1965a, Nature, 206:338.
Du Praw, E.J., 1965b, Proc. Nat. Acad. Sci. USA., 53:161.
Du Praw, E.J., and P.M.M. Rae, 1966, Nature, 212:598.
Ebel-Tsipsis, J., D. Botstein, and M.S. Fox, 1972, J. Mol. Biol., 71:449.
Echols, H., L. Moore, and R. Gingery, 1968, J. Mol. Biol., 34:251.
Edelman, G.M., and J.A. Gally, 1967, Proc. Nat. Acad. Sci. USA., 57:354.
Edgar, R.S., 1958, Genetics, 43:235.
Edgar, R.S., 1961, Virology, 13:1.
Edgar, R. S., R. P. Feynman, S. Klein, J. Lielausis and C.M. Steinberg, 1962, Genetics, 47:179.
Efremova, G.I., 1968, Genetika, 9:47.
Ellingboe, A.H., 1963, Proc. Nat. Acad. Sci. USA., 49:286.
Emerson, R.A., and M. Rhoades, 1933, Amer. Naturalist, 67:374.
Emerson, S., 1956, Compt. Rend. Trav. Lab. Carlsberg, Sér. Physiol., 26:71.
Emerson, S., 1963, in: Methodology in Basic Genetics, San Francisco, p. 289.
Emerson, S., 1966a, Genetics, 53:475.
Emerson, S., 1966b, in: The Fungi, 2, New York–London, p. 513.
Emerson, S., 1969, in: Genetic Organization, 1, New York–London, p. 267.
Emerson, S., and C.C.C. Yu-Sun, 1967, Genetics, 55:39.
Emmerson, P.T., 1968, Genetics, 60:19.
Ende, van den, P., and N. Symonds, 1972, Molec. Gen. Genet. 116:239.
Engelhardt, P., and K. Pusa, 1972, Nature New Biol. 240:163.
Enns, H., and E.N. Larter, 1963, Canad. J. Genet. Cytol., 4:263.
Enquist, L.W., and A. Skalka, 1973, J. Mol. Biol., 75:185.
Ephrati-Elizur, E., P.R. Srinivasan, and S. Zamenhof, 1961, Proc. Nat. Acad. Sci., USA., 47:56.
Ephrussi-Taylor, H., 1961, in: Growth in the Living Systems, New York, p. 39.
Ephrussi-Taylor, H., and T.C. Gray, 1966, J. Gen. Physiol., 49, 9, Part 2:1372.
Ephrussi-Taylor, H., A.M. Sicard, and R. Kamen, 1965, Genetics, 51:455.
Esposito, M.S., 1968, Genetics, 58:507.
Esposito, M.S., 1971, Molec. Gen. Genet. 111:297.
Esposito, R.E., 1968, Genetics, 59:191.
Evans, G.M., and A.J. Macefield, 1972, Nature New Biol., 236:110.
Evans, H.J., 1962, Internat. Rev. Cytol., 13:222.
Evans, H.J., 1966, in: Genetical Aspects of Radiosensitivity: Mechanism of Repair, Vienna, p. 31.

Fabergé, A.C., 1942, J. Genet., 43:121.
Fabre, F., 1971., Molec. Gen. Genet., 110:134.
Fabre, F., 1972, Molec. Gen. Genet., 117:153.
Fan, D.P., 1969, Genetics, 61:351.
Fareed, G.C., and C.C. Richardson, 1967, Proc. Nat. Acad. Sci. USA., 58:665.
Fiandt, M., Z. Hradecna, H.A. Lozeron, and W. Szybalski, 1971, in: The Bacteriophage Lambda, Cold Spring Harbor Lab., p. 329.
Fields, K., and L.S. Olive, 1967, Genetics, 57:483.
Fincham, J.R.S., 1966, Genetic Complementation, Benjamin, New York and Amsterdam. (Russian translation, Moscow, 1968).
Fincham, J.R.S., 1967, Genet. Res., 9:49.
Fincham, J.R., and R. Holliday, 1970, Mol. Gen. Genetics, 109:309.
Fisher, K.M., and H. Bernstein, 1965, Genetics, 52:1127.
Fogel, S., and D.D. Hurst, 1963, Genetics, 48:321.
Fogel, S., and D.D. Hurst, 1967, Genetics, 57:455.
Fogel, S., and R.K. Mortimer, 1968, Genetics, 60, 1, Part 2, p. 178.
Fogel, S., and R.K. Mortimer, 1969, Proc. Nat. Acad. Sci. USA., 62:96.
Fogel, S., and R.K. Mortimer, 1970, Mol. Gen. Genetics, 109:177.
Fogel, S., D.D. Hurst, and R.K. Mortimer, 1970, Genetics, 64, suppl. p. 21.
Folsome, C.E., 1962, Z. Vererbungsl., 93:404.
Folsome, C.E., 1965, Genetics, 51:391.
Fortuin, J.J.H., 1971, Mut. Res., 13:137.
Foss, H.M., and F.W. Stahl, 1963, Genetics, 48:1659.
Fox, M.S., 1966, J. Gen. Physiol., 49, 9, Part 2, p. 183.
Fox, M.S., and M.K. Allen, 1964, Proc. Nat. Acad. Sci. USA., 52:412.
Franclin, N.C., 1967, Genetics, 55:699.
Franklin, N.G., 1971a, in: The Bacteriophage Lambda, Cold Spring Harbor Lab., p. 175.
Franklin, N.G., 1971b, in: The Bacteriophage Lambda, Cold Spring Harbor Lab., p. 621.
Frédéricq, P., 1972, Lunteren Lectures on Molecular Genetics, S4:1.
Freese, E., 1957, Genetics, 42:677.
Freese, E., 1958, Cold Spring Harbor Sympos. Quant. Biol., 23:13.
Freese, E., and H.B. Starck, 1962, Proc. Nat. Acad. Sci. USA., 48:1796.
Freese, E.B., and E. Freese, 1966, Genetics, 54:1055.
Friedberg, E.C., 1972, Mut. Res., 15:113.
Friedberg, E.C., and D.A. Goldthwait, 1968, Cold Spring Harbor Sympos. Quant. Biol., 33:271.
Friedman, E., and H.O. Smith, 1972, J. Biol. Chem., 247:2846.
Friedman, E., and H.O. Smith, 1973, Nature New Biol. 24:54.
Friedrich-Freksa, H., 1940, Naturwiss., 28:376.
Frost, L.C., 1961, Genet. Res., 2:43.
Gabor, M., and R.D. Hotchkiss, 1966, Proc. Nat. Acad. Sci. USA., 56:1441.
Gajewski, W., A. Makarewicz, A. Kruszewska, A. Paszewski, and S. Surzycki, 1966, in: The Physiology of Gene and Mutation Expression, Prague, p. 75.
Gajewski, W., A. Paszewski, A. Dawidowicz, and B. Dudzinska, 1968, Genet. Res., 11:311.
Gall, J.G., 1956, Brookhaven Symp. Biol., 8:17.
Gall, J., 1963a, Science, 139:120.
Gall, J., 1963b, Nature, 198:36.
Gall, J., 1966, Chromosoma, 20:221.
Gallant, J., and T. Spotswood, 1965, Genetics, 52:107.
Ganesan, A.T., and N. Buckman, 1968, Biophys. J., 9, Abstr., A18.
Gassner, G., 1967, J. Cell Biol., 35, 2, Part 2, p. 166A.

Gatti, M., and G. Olivieri, 1973, Mut. Res., 17:101.
Gefter, M.L., Y. Hirota, T. Kornberg, J.A. Wechsler, and C. Barnoux, 1971, Proc. Nat. Acad. Sci. USA., 68:3150.
Gellert, M., 1967, Proc. Nat. Acad. Sci. USA., 57:148.
George, D., and D. Rosenberg, 1972, J. Bacteriol., 112:1017.
Giles, N.H., 1951, Cold Spring Harbor Sympos. Quant. Biol., 16:283.
Giles, N.H., F.J. de Serres, and E. Barbour, 1957, Genetics, 42:608.
Gillies, C.B., 1972, Chromosoma, 36:119.
Glass, B., 1957, in: The Chemical Basis of Heredity, Baltimore, p. 257.
Goldfarb, D.M., G.I. Goldberg, L.S. Chernin, L.A. Gukova, I.D. Avdienko, B.N. Kuznetsova, and I.C. Kushner, 1973, Molec. Gen. Genet., 120:211.
Goldman, S.L., and H. Gutz, 1971, Genetics, 69 Abstracts, 23.
Goldmark, P.J., and S. Linn, 1970, Proc. Nat. Acad. Sci. USA., 67:434.
Goldmark, P.J., and S. Linn, 1972, J. Biol. Chem., 247:1849.
Goodgal, S.H., and N. Notani, 1968, J. Mol. Biol., 35:449.
Goodgal, S.H., and E. Postel, 1967, J. Mol. Biol., 28:261.
Gottesman, M.E., and R. A. Weisberg, 1971, in: The Bacteriophage Lambda, Cold Spring Harbor Lab., p. 113.
Gowen, I.W., 1933, J. Exper. Zool, 65:83.
Grace, D., 1970, Mutat. Res., 10:489.
Gray, T.C., and H. Ephrussi-Taylor, 1967, Genetics, 57:125.
Green, M.M., and K.C. Green, 1949, Proc. Nat. Acad. Sci. USA., 35:586.
Greenblatt, J.M., 1968, Genetics, 58:585.
Grell, E.H., 1964, Genetics, 50: Abstr. p. 251.
Grell, R.F., 1966, Genetics, 54:411.
Grell, R.F., 1973, Genetics, 73:87.
Grell, R.F., and A.C. Chandley, 1965, Proc. Nat. Acad. Sci. USA., 53:1340.
Grell, R.F., H. Bank, and G. Gassner, 1972, Nature New Biol., 240:155.
Gross, J., and E. Englesberg, 1959, Virology, 9:314.
Gross, J., and M. Gross, 1969, Nature, 224:1166.
Grossman, L., J.C. Kaplan, S.R. Kushner, and J. Mahler, 1968. Cold Spring Harbor Sympos. Quant. Biol., 33:229.
Guarneros, G., and H. Echols, 1970, J. Mol. Biol., 47:565.
Guerola, N., J.L. Ingraham, and E. Cerdá-Olmedo, 1971, Nature New Biol., 230:122.
Guerrini, F., and M.S. Fox, 1968a, Proc. Nat. Acad. Sci. USA., 59:429.
Guerrini, F., and M.S. Fox, 1968b, Proc. Nat. Acad. Sci. USA., 59:1116.
Guest, J.R., and C. Yanofsky, 1966, Nature, 210:799.
Guglielminetti, R., 1968, Mutat. Res., 5:225.
Gurney, T., and M.S. Fox, 1968, J. Mol. Biol., 32:83.
Gussin, G.N., and V. Peterson, 1972, J. Virology, 10:760.
Gutz, H., 1968, Proc. 12th Internat. Congr. Genet., part I (1.2.9), Tokyo.
Gutz, H., 1971a, Genet. Res., 17:45.
Gutz, H., 1971b, Genetics, 69:317.
Gutz, H., 1971c, Genetics, 69 Abstracts, s26.
Haan, P.G. de, W.P. Hoekstra, and C. Verhoef, 1972, Mut. Res., 14:375.
Haberman, A., J. Heywood, and M. Meselson, 1972, Proc. Nat. Acad. Sci. USA., 69:3138.
Haefner, K., 1967, Genetics, 57:169.
Haefner, K., and L. Howrey, 1967, Mutat. Res., 4:219.
Hagemann, R., 1958, Z. Vererbungsl., 89:587.
Hagemann, R., and B. Snoad, 1971, Heredity, 27:409.
Haldane, J.B.S., 1919, J. Genetics, 8:291.
Haldane, J.B.S., 1931, Cytologia, 3:54.

BIBLIOGRAPHY

Hall, J.C., 1972, Genetics, 71:367.
Hall, J.D., and P. Howard-Flanders, 1972, J. Bacteriol. 110:578.
Hammerl, H. and W. Klingmuller, 1972, Z. Naturforsch., 27b:68.
Hanawalt, P.C., D.E. Pettijohn, E.C. Pauling, C.F. Brunk, D.W. Smith, L.C. Kanner, and J.L. Couch, 1968, Cold Spring Habor Sympos. Quant. Biol., 33:187.
Harm, W., 1964, Mutat. Res., 1:344.
Harris, W.J., and G.C. Barr, 1969, J. Mol. Biol., 39:245.
Harris, W.J., and G.C. Barr, 1971a, Molec. Gen. Genet., 113:316.
Harris, W.J., and G.C. Barr, 1971b, Molec. Gen. Genet., 113:331.
Hastings, P.J., 1972, Genet. Res., 20:253.
Hastings, P.J., and H.L.K. Whitehouse, 1964, Nature, 201:1052.
Hayes, W., 1968, The Genetics of Bacteria and Their Viruses, Second Edition, Blackwell, Oxford. (Russian translation of First Edition, Moscow, 1965).
Haynes, R.H., 1966, Radiat. Res., 6, Suppl. 1, p. 1.
Heddle, I.A., 1969, J. Theoret. Biol., 22:151.
Hedgpeth, J., H.M. Goodman, and H.W. Boyer, 1972, Proc. Nat. Acad. Sci. USA., 69:3448.
Helinski, D.R., and C. Yanofsky, 1962, Proc. Nat. Acad. Sci. USA., 48:173.
Helling, R.B., 1967, Genetics, 57:665.
Henderson, S.A., 1964, Chromosoma, 15:345.
Henderson, S.A., 1966, Nature, 211:1043.
Henderson, S.A., 1972, Chromosoma, 35:41.
Herman, R.K., 1965, J. Bacteriol., 90:166.
Herman, R.K., 1968a, J. Bacteriol., 96:173.
Herman, R.K., 1968b, Genetics, 58:55.
Hershey, A.D., 1952, Internat. Rev. Cytol., 1:119.
Hershey, A.D., 1958, Cold Spring Harbor Sympos. Quant. Biol., 23:19.
Hershey, A.D., and M. Chase, 1951, Cold Spring Harbor Sympos. Quant. Biol., 16:471.
Hershey, A.D., and R. Rotman, 1949, Genetics, 34:44.
Hertel, R., 1963, Z. Vererbungsl., 94:436.
Hertman, J.M., 1967, J. Bacteriol., 93:580.
Hexter, W.M., 1963, Proc. Nat. Acad. Sci. USA., 50:372.
Hinton, C.W., 1970, Genetics, 66:663.
Hirota, Y., M. Gelfter, and L. Mindich, 1972, Proc. Nat. Acad. Sci. USA., 69:3238.
Hobom, B., and G. Hobom, 1972, Molec. Gen. Genet., 117:229.
Holland, I.B., and E.J. Threlfall, 1969, J. Bacteriol., 97:91.
Holliday, R., 1962, Genet. Res., 3:472.
Holliday, R., 1964a, Genetics, 50:323.
Holliday, R., 1964b, Genet. Res., 5:282.
Holliday, R., 1965, Genet. Res., 6:104.
Holliday, R., 1966a, Genet. Res., 8:323.
Holliday, R., 1966b, Heredity, 21:339.
Holliday, R., 1968, in: Replication and Recombination of the Genetic Material, Canberra.
Holliday, R., 1970, Prokaryotic and Eukaryotic Cells, London, p. 359.
Holliday, R., 1971, Nature New Biol., 232:233.
Holliday, R., and W.K. Hollaman, 1972, Lunteren Lectures on Molecuar Genetics, Vol. 4.
Holliday, R., and H.L.K. Whitehouse, 1970, Mol. Gen. Genetics, 107:85.
Holloway, B.W., 1966, Mutat. Res., 3:452.
Honda, M., and H. Uchida, 1969, Genetics, 63:743.
Hopwood, D.A., 1967, Bacteriol. Rev., 31:373.
Horowitz, N.H., 1957, in: The Chemical Basis of Heredity, Baltimore, p. 3.
Hosoda, J., and E. Matheus, 1968, Proc. Nat. Acad. Sci. USA., 61:997.

Hotchkiss, R.D., 1958, Symp. Exper. Biol., 12:49.
Hotchkiss, R.D., 1971, Advances in Genetics, 16:327.
Hotchkiss, R.D., and A.H. Evans, 1958, Cold Spring Harbor Sympos. Quant. Biol., 23:85.
Hotta, Y., and H. Stern, 1971a, Nature New Biol., 234:83.
Hotta, Y., and H. Stern, 1971b, J. Mol. Biol., 55:337.
Hotta, Y., M. Ito and H. Stern, 1966, Proc. Nat. Acad. Sci. USA., 56:1184.
Howard-Flanders, P., 1967, J. Cellular and Compar. Physiol., 70, Suppl. 1, p. 73.
Howard-Flanders, P., and R.P. Boyce, 1964, Genetics, 50:256.
Howard-Flanders, P., and R.P. Boyce, 1966, Radiat. Res., 6, Suppl., p. 156.
Howard-Flanders, P., R.P. Boyce, E. Simons and L. Theriot, 1962, Proc. Nat. Acad. Sci. USA., 48:2109.
Howard-Flanders, P., and L. Theriot, 1966, Genetics, 53:1137.
Howard-Flanders, P., B.M. Wilkins and W.D. Rupp, 1968, in: Molecular Genetics, Berlin–New York, p. 113.
Howe, H.B., 1956, Genetics, 41:610.
Howell, S.H., and H. Stern, 1971, J. Mol. Biol., 55:357.
Hsu, T.C., 1964, J. Cell Biol., 23:53.
Huberman, J.A., and A.D. Riggs, 1966, Proc. Nat. Acad. Sci. USA., 55:599.
Hunnable, E.G., and B.S. Cox, 1971, Mut. Res., 13:297.
Hurst, D.D., and S. Fogel, 1964, Genetics, 50:435.
Hurwitz, J., 1967, J. Cellular and Compar. Physiol., 70, Suppl. 1, p. 211.
Huskins, L., and S.G. Smith, 1935, Amer. J. Bot., 49:119.
Huskins, L., 1952, Internat. Rev. Cytol., 1:9.
Hüsslein, V., B. Otto, F. Bonhoeffer, and H. Schaller, 1971, Nature New Biol. 234:285.
Iha, K.K., 1967, Genetics, 57:865.
Iha, K.K., 1969, Mol. Gen. Genetics, 105:30.
Ihler, G., and M. Meselson, 1963, Virology, 21:7.
Ishikawa, T., 1962, Genetics, 47:1147.
Ito, M., H. Stern and Y. Hotta, 1967, Developm. Biol., 16:54.
Iyengar, N.K., 1939, Annual Rev. Bot., 3:271.
Iyer, V.N., and W.D. Rupp, 1971, Biochim. Biophys. Acta, 228:117.
Jacob, F., 1965, Genetika, 6:3.
Jacob, F., and E.L. Wollman, 1955, Ann. Inst. Pasteur, 88: 724.
Jacob, F., and E.L. Wollman, 1961, Sexuality and the Genetics of Bacteria, Academic Press, New York. (Russian translation, Moscow, 1962).
Jacobs, A., A. Bopp, and U. Hagen, 1972, Int. J. Rad. Biol., 22:431.
James, A.P., and M.M. Werner, 1966, Mutat. Res., 3:477.
Jansen, G.J.O., 1970, Mutat. Res., 10:33.
Janssens, F.A., 1909, Cellule, 25:387.
Jessop, A.P., and D.G. Catcheside, 1965, Heredity, 20:237.
John, B., and K.R. Lewis, 1965, Protoplasmatologia, 6:1.
Jones, G.H., and R.J. Brumpton, 1971, Chromosoma, 33:115.
Jordan, E., and M. Meselson, 1965, Genetics, 51:77.
Jordan, P., 1941, Naturwiss, 29:89.
Judd, B.H., M.W. Shen and T.C, Kaufman, 1972, Genetics, 71:139.
Kaiser, A.D., 1955, Virology, 1:424.
Kakar, S.N., 1963, Genetics, 48:957.
Kaplan, J.C., S.R. Kushner, and L. Grossman, 1969, Proc. Nat. Acad. Sci. USA., 63:144.
Kaplan, S., J. Suyama, and D.M. Bonner, 1964, Genetics, 49:145.
Kapp, D.S., and K.C. Smith, 1970, J. Bacteriol., 103:49.
Kaufmann, B., 1957, Bot. Rev., 14:48.
Kaufmann, B., H. Gay, and M.J. MacElderry, 1957, Proc. Nat. Acad. Sci. USA., 43:255.

Kavenoff, R., and B.H. Zimm, 1973, Chromosoma, 41:1.
Kawai, S., and H. Hanafusa, 1972, Virology, 49:37.
Kayajanian, G., 1972, Virology, 49:599.
Kellenberger, G., M.L. Zichichi, and J. Weigle, 1961, Proc. Nat. Acad. Sci. USA., 47:869.
Kellenberger, G., M.L. Zichichi, and H.L. Epstein, 1962, Virology, 17:44.
Kellenberger-Gujer, R., 1971, in: The Bacteriophage Lambda, Cold Spring Harbor Lab., p. 417.
Kellenberger-Gujer, G., and R.A. Weisberg, 1971, in: The Bacteriophage Lambda, Cold Spring Harbor Lab., p. 407.
Kelly, R.B., M.R. Atkinson, J.A., Huberman, and A. Kornberg, 1969, Nature, 224:495.
Kelly, S.W., and H.J. Whitfield, Jr., 1971, Nature, 230:33.
Kelly, T.J., and H.O. Smith, 1970, J. Mol. Biol. 51:393.
Kemper, B., 1970, Mol. Gen. Genetics, 107:107.
Kidwell, M.G., 1972, Genetics, 70:419.
Kihlman, B.A., and B. Hartley, 1967, Hereditas, 57:289.
Kikkawa, H., 1934, J. Genetics, 28:329.
Kimball, R.F., 1964, Mutat. Res., 1:129.
King, R.C., 1970, Internat. Rev. Cytol., 28:125.
Kitani, Y., 1972, Genetics 71 Abstracts, s30.
Kitani, Y., and L.S. Olive, 1967, Genetics, 57:767.
Kitani, Y., and L.S. Olive, 1969, Genetics, 62:23.
Kitani, Y., and L. Olive, 1970, Proc. Nat. Acad. Sci. USA., 66:1290.
Kitani, Y., L.S. Olive, and A.S. El-Ani, 1962, Amer. J. Bot., 49:697.
Kleinschmidt, A.K., 1967, in: Molecular Genetics, 2, New York–London, p.47.
Kleinschmidt, A.K., D. Lang, D. Jacherts, and K.K. Zahn, 1962, Biochim. Biophys. Acta, 61:857.
Kniep, H., 1928, Die Sexualität der niederen Pflanzen, Jena.
Kohoutova, M., I. Holubova, H. Brana, and P. Tichy, 1970, Folia Microbiol., 15:183.
Kölsch, E., and P. Starlinger, 1965, Z. Vererbungsl., 96:297.
Kornberg, A., 1961, Enzymatic Synthesis of DNA, London.
Kornberg, A., 1969, Science, 163:1410.
Korogodin, V.A., 1966, Problems in Postradiation Recovery [in Russian], Moscow.
Korolev, V.G., and L.M. Gracheva, 1972, Genetika, 8(8):111.
Kosambi, D.D., 1944, Ann. Eugenics, 12:172.
Kozhina, T.N., 1968a, Genetika, 9:65.
Kozhina, T.N., 1968b, Genetika, 12:36.
Kozinski, A.W., 1961, Virology, 13:124.
Kozinski, A.W., 1968, Cold Spring Harbor Sympos. Quant. Biol., 33:375.
Kozinski, A.W., and Z.Z. Felgenhauer, 1967, J. Virology, 1:1193.
Kozinski, A.W., and P.B. Kozinski, 1964, Proc. Nat. Acad. Sci. USA., 52:211.
Kozinski, A.W., and P.B. Kozinski, 1969, J. Virology, 3:85.
Kozinski, A.W., P.B. Kozinski, and P. Shannon, 1963, Proc. Nat. Acad. Sci. USA., 50:746.
Krish, H.M., N.V. Hamlett, and H. Berger, 1972, Genetics, 72: 187.
Kriviskii, A.S., 1966, in: The Physiology of Gene and Mutation Expression, Prague, p. 75.
Kruszewska, A., and W. Gajewski, 1967, Genet. Res., 9:159.
Krylov, V.N., 1972, Virology, 50:291.
Krylov, V.N., and S.I. Alikhanyan, 1967, Genetika, 5:128.
Kubitschek, H.E., 1966, Proc. Nat. Acad. Sci. USA., 55:265.
Kunicki-Goldfinger, W.J.H., 1968, Acta Microbiol. Polon., 17:147.
Kushev, V.V., 1970, in: Molecular Mechanisms of Genetic Processes [in Russian], Moscow, p. 67.

Kushev, V.V., S.E. Bresler, and R.A. Kreneva, 1970, in: Synthesis, Structure, and Properties of Polymers [in Russian], Leningrad, p. 245.
Kushner, S.R., J.C. Kaplan, H. Ono, and L. Grossman, 1971a, Biochemistry, 10:3325.
Kushner, S. R., H. Nagaishi, A. Templin, and A. J. Clark, 1971b, Proc. Nat. Acad. Sci. USA., 68:824.
Kushner, S.R., H. Nagaishi, and A.J. Clark, 1972, Proc. Nat. Acad. Sci. USA., 68:1366.
Kutter, E.M., and J.S. Wiberg, 1968, J. Mol. Biol., 38:395.
Lacks, S., 1966, Genetics, 53:207.
Lacks, S., 1970, J. Bacteriol., 101:373.
Lacks, S., and R.D. Hotchkiss, 1960, Biochim. et Biophys. Acta, 39:508.
LaCour, L.E., and B. Wells, 1970, Chromosoma, 29:419.
Laipis, P.J., and A.T. Ganesan, 1972, Proc. Nat. Acad. Sci. USA., 69:3211.
Laird, C., L. Wang and W. Bodmer, 1968, Mutat. Res., 6:205.
Lamb, B.C., 1969, Genetics, 63:807.
Lamb, B.C., 1971, Genet. Res., 18:255.
Lambert, B., 1972, J. Mol. Biol., 72:65.
Landner, L., 1972, Molec. Gen. Genet., 119:103.
Lavigne, S., and L.C. Frost, 1964, Genet. Res., 5:366.
Lawrence, C.W., 1961a, Heredity, 16:83.
Lawrence, C.W., 1961b, Radiat. Bot., 1:92.
Lawrence, C.W., 1965, Nature, 206:789.
Lawrence, C.W., 1967, Genet. Res., 9:123.
Lawrence, C.W., 1968, Heredity, 23:143.
Lawrence, C.W., 1970, Mutat. Res., 10:557.
Lawrence, M.J., 1963, Heredity, 18:27.
Lawrence, P. St., 1956, Proc. Nat. Acad. Sci. USA., 42:189.
Lea, D.E., 1955, Actions of Radiations on Living Cells, Second Edition, Cambridge.
Leavitt, R.W., J. Wohlhieter, E.M. Johnson, G.E. Olson, and L.S. Baron, 1971, J. Bacteriol., 108:1357.
Leblon, G., 1972a, Molec. Gen. Genet., 115:36.
Leblon, G., 1972b, Molec. Gen. Genet., 116:322.
Leblon, G., and J.-L. Rossignol., 1973, Molec. Gen. Genet., 122:165.
Leclerc, J.E., and J.K. Setlow, 1972, J. Bacteriol., 110:930.
Lederberg, J., 1947, Genetics, 32:505.
Lederberg, J., 1955, J. Cellular and Compar. Physiol., 45, Suppl. 2:75.
Lederberg, J., 1957, Proc. Nat. Acad. Sci. USA., 42:1060.
Lee, W.R., G.A. Sega and C.F. Alford, 1967, Proc. Nat. Acad. Sci. USA., 58:1472.
Lefevre, J.G., 1971, Genetics, 67:497.
Lehmann, A.R., 1972a, Biophys. J., 12:1316.
Lehmann, A.R., 1972b, J. Mol. Biol., 66:319.
Lehmann, A.R., 1972c, Europ. J. Biochem., 31:438.
Lehman, I. R., 1966, in: The Nucleic Acids (Russian translation, Moscow, p. 97).
Levinthal, C., 1954, Genetics, 39:169.
Levinthal, C., 1956, Proc. Nat. Acad. Sci. USA., 42:394.
Levinthal, C., 1959, in: The Viruses, 2, New York, p. 281.
Lewis, E.B., 1941, Proc. Nat. Acad. Sci. USA., 27:31.
Lewis, E.B., 1945, Genetics, 30:137.
Lewis, E.B., 1951, Cold Spring Harbor Sympos. Quant. Biol., 16:159.
Lewis, E.B., 1952, Proc. Nat. Acad. Sci. USA., 38:953.
Lewis, E.B., 1967, in: Heritage from Mendel, Madison, p. 17.
Lewis, L.A., 1972, Lunteren Lectures on Molecular Genetics, S9:1.

BIBLIOGRAPHY

Lima de Faria A. (Ed)., 1969, Handbook of Molecular Cytology, Amsterdam.
Lima de Faria, A. and H. Jaworska, 1972, Hereditas, 70:39.
Lindahl, G., 1969, Virology, 39:839, 861.
Lindegren, C.C., 1933, Bull. Torrey Bot. Club, 60:133.
Lindegren, C.C., 1949, The Yeast Cell, its Genetics and Cytology. Saint Louis.
Lindegren, C.C., 1953, J. Genetics, 51:625.
Lindegren, C.C., 1955, Science, 121:605.
Lindegren, C.C., 1964, Nature, 204:322.
Lindegren, C.C., and G. Lindegren, 1937, J. Heredity, 3:105.
Lindegren, C.C., and G. Lindegren, 1942, Genetics, 27:1.
Lissouba, P., 1960, Ann. Sci. Natur. Bot. et Biol. Végét., 44:641.
Lissouba, P., J. Mousseau, G. Rizet, and J.L. Rossignol., 1962, Advances in Genetics, 11:343.
Lissouba, P., and G. Rizet., 1960, C. r. Acad. Sci., Sér. D, 250:3408.
Lloyd, R.G., 1972, Lunteren Lectures on Molecular Genetics, S2:4.
Lobashev, M.E., 1967, Genetics [in Russian], Second Edition, Leningrad.
Lock, R.H., 1906, Recent Progress in the Study of Variation, Heredity and Evolution, London.
Lotan, D., E. Yagil, and M. Bracha, 1972, Genetics, 72:381.
Low, B., 1968, Proc. Nat. Acad. Sci. USA., 60:160.
Low, B., 1973, Molec. Gen. Genet., 122:119.
Louarn, J.H., and A.M. Sicard, 1968a, Biochem. Biophys. Res. Commun., 30:683.
Louarn, J. H., and A.M. Sicard, 1968b, Biochem. Biophys. Res. Commun., 32:461.
Louarn, J.H., and A.M. Sicard, 1969, Biochem. Biophys. Res. Commun., 36:101.
Lu, B.C., 1966, Exper. Cell Res., 43:224.
Lu, B.C., 1967, J. Cell Sci., 2:529.
Lucchesi, J.C., and D.T. Suzuki, 1968, Annual Rev. Genetics, 2:53.
Luchnik, N.V., N.A. Poryadkova, L.S. Tsarapkin, an N. V. Timofeev-Resovskii, 1964, in: Repair Processes after Radiation Injuries [in Russian], Moscow, p. 5.
Lucia, P., de and J. Cairns, 1969, Nature, 224:1164.
Luzzati, M., L. Clavilier, and P.P. Slonimski, 1959, Cr. r. Acad. Sci., Sér. D, 249:1412.
Lyons, L.B., and N. Zinder, 1972, Virology, 49:45.
Maccacaro, G.A., and W. Hayes, 1961, Genet. Res., 2:406.
Mackendrich, M.E., and G. Pontecorvo, 1952, Experientia, 8:390.
Maeda, T., 1930, Japan, J. Genet., 15:118.
Magni, G.E., and R.C. von Borstel, 1962, Genetics, 47:1097.
Mahler, J., S.R. Kushner, and L. Grossman, 1971, Nature New Biol., 234:47.
Makarewicz, A., 1964, Acta Soc. Bot. Polon., 33:1.
Mandel, M., and B. Kornreich, 1972, Virology, 49:300.
Manney, T.R., and R.K. Mortimer, 1964, Science, 143:581.
Maquire, M.P., 1966a, Genetics, 53:1071.
Maquire, M.P., 1966b, Proc. Nat. Acad. Sci. USA., 55:44.
Maquire, M.P., 1967, Chromosoma, 21:221.
Maquire, M.P., 1968a, Proc. Nat. Acad. Sci. USA., 60:533.
Maquire, M.P., 1968b, Genetics, 60:353.
Maquire, M.P., 1968c, Genet. Res., 12:21.
Maquire, M.P., 1969, Nature, 222:691.
Maguire, M.P., 1972, Genetics, 70:355.
Marcou, D., J. Mousseau, G. Rizet, and J.L. Rossignol, 1965, Ann. Genet., 6, 2:113.
Margolin, P., and F.H. Mukai, 1966, Proc. Nat. Acad. Sci. USA., 55:283.
Mario, A., 1966, Atti Acad. Naz. Lincei, Mem. Cl. Sci. Fis., Mat. e Natur., 40:988.
Marmur, J., and D. Lane, 1960, Proc. Nat. Acad. Sci. USA., 46:453.

Martin, G., G. Ketele, and J. Guillaume, 1969, Compt. Rend. Acad. Sci., Paris, ser. D, 268:609.
Masker, W.E., and P.C. Hanawalt, 1973, Proc. Nat. Acad. Sci. USA., 70:129.
Mather, K., 1933a, J. Genetics, 27:243.
Mather, K., 1933b, Amer. Naturalist, 67:354.
Mather, K., 1936a, J. Genetics, 32:287.
Mather, K., 1936b, J. Genetics, 33:207.
Mather, K., 1937, Cytologia (Jub. vol), p. 514.
Mather, K., 1938, Biol. Rev. Cambridge Philos. Soc., 13:252.
Mather, K., 1951, The Measurement of Linkage in Heredity, London.
Matney, T.S., B.V. Nguyen, M.A. Butler, and E.P. Goldschmidt, 1971, Genetics 69 Abstracts, s42.
Matsuura, H., 1940, Cytologia, 10:390.
Mattem, I.E., 1972, Lunteren Lectures on Molecular Genetics, S2:1.
Matvienko, N.I., 1968, Dokl. Akad. Nauk SSSR, 179:1457.
Matvienko, N.I., 1969, Genetika, 5:84.
Matvienko, N.I., 1972a, Molec. Gen. Genet., 117:45.
Matvienko, N.I., 1972b, Genetika, 8(9):119.
Maworth, J.W., and H.K. Swenson, 1923, Science, 58:124.
Maynard-Smith, S., and N. Symonds, 1973, J. Mol. Biol., 74:33.
Mazza, G., H.M. Eisenstark, M.C. Serra, and M. Polisinelli, 1972, Molec. Gen. Genet., 115:73.
McClintock, B., 1945, Amer. J. Bot., 32:671.
McGrath, R.A., and R.W. Williams, 1966, Nature, 212:534.
McMilin, K.D., and V.E.A. Russo, 1972, J. Mol. Biol., 68:49.
McNelly-Ingle, C.A., B.C. Lamb, and L.C. Frost, 1966, Genet. Res., 7:169.
Mead, C.G., 1964, Proc. Nat. Acad. Sci. USA., 52:1482.
Melechen, N.E., and T.A. Hudnik-Plevnik, 1972, Proc. Nat. Acad. Sci. USA., 69:3195.
Mendel, G., 1866, Verhandl. Naturforsch., Vereines Brunn, 4:3.
Meselson, M., 1964, J. Mol. Biol., 9:734.
Meselson, M., 1965, in: Heritage from Mendel, Madison, p. 81.
Meselson, M., 1967, J. Cell. Comp. Physiol., 7, Suppl. 1:113.
Meselson, M., 1972, J. Mol. Biol., 71:795.
Meselson, M., and F. Stahl, 1958, Proc. Nat. Acad. Sci. USA., 44:671.
Meselson, M., and J.J. Weigle, 1961, Proc. Nat. Acad. Sci. USA., 47:857.
Meyer, G.F., 1960, Proc. 2nd Europ. Regional Conf. Electron Microscopy, 2, Delft, p. 951.
Meyer, G.F., 1964, Proc. 3rd Europ. Regional Conf. Electron Microscopy, 2, Prague, p. 461.
Michalke, W., 1967, Mol. Gen. Genetics, 99:12.
Miller, O.L., 1964, J. Cell Biol., 23:109A.
Miller, O.L. Jr., and B.A. Hamkalo, 1972, Internat. Rev. Cytol., 33:1.
Millington-Ward, A.M., 1969, Genetica, 40:339.
Mirsky, A.E., 1971, Proc. Nat. Acad. Sci. USA., 68:2945.
Mitchell, H., 1957, in: The Chemical Basis of Heredity, Baltimore, p. 79.
Mitchell, M.B., 1955, Proc. Nat. Acad. Sci. USA., 41:215.
Mitchell, M.B., 1956, Compt. Rend. Trav. Lab. Carlsberg, Sér. Physiol., 26:285.
Mitchell, M.B., 1960, Genetics, 45:507.
Mitra, S., 1958, Genetics, 43:771.
Modigliani, P., 1967, Genetics, 56:7.
Moens, P.B., 1964, Chromosoma, 15:231.
Moens, P.B., 1968, Genetics, 60, Part 2, p. 205.
Moens, P.B., 1969, Chromosoma, 28:1.

Moens, P.B., 1970, Proc. Nat. Acad. Sci. USA., 66:94.
Monesi, V., M. Crippa and R. Zito-Bignamy, 1967, Chromosoma, 21:369.
Monk, M., M. Peacey, and J.D. Gross, 1971, J. Mol. Biol., 58:623.
Moody, E.E.M., K.B. Low, and D.W. Mount, 1973, Molec, Gen. Genet., 121:197.
Moore, D., 1972, Genet. Res., 19:281.
Morgan, T.H., 1911a, J. Exper. Zool., 11:365.
Morgan, T.H., 1911b, Amer. Naturalist, 45:65.
Morgan, T.H., 1911c, Science, 34:384.
Morgan, T.H., 1919, The Physical Basis of Heredity, Philadelphia.
Morgan, T.H., 1926, The Theory of the Gene, New Haven. (Russian translation, Moscow, 1927).
Morgan, T.H., and E. Cattell, 1912, J. Exper. Zool., 13:79.
Morgan, T.H., A.H. Sturtevant, C.B. Bridges, J. Schultz, and H.J. Muller, 1915, The Mechanism of Mendelian Heredity, New York.
Mori, S., and S. Nakai, 1972, Molec. Gen. Genet., 117:187.
Morpurgo, G., and L. Volterra, 1968, Genetics, 58:529.
Morrison, D.A., and W.R. Guild, 1972a, J. Bacteriol., 112:220.
Morrison, D.A., and W.R. Guild, 1972b, J. Bacteriol., 112:1157.
Morse, H.G., and L.S. Lerman, 1969, Genetics, 61:41.
Mortelmans, K., and E.C. Friedberg, 1972, J. Virol., 10:730.
Moseley, B.E.B., 1969, J. Bacteriol., 97:647.
Moses, M.J., 1964, in: Cytology and Cell Physiology, New York–London, p. 423.
Moses, M.J., 1956, J. Biophys. Biochem. Cytol., 2:215.
Moses, M.J., 1968, Annual Rev. Genetics, 2:363.
Moses, M.J., and J.H. Taylor, 1955, Exper. Cell Res., 9:474.
Moses, R.F., and C.C. Richardson, 1970, Biochem. Biophys. Res. Commun., 41:1557.
Mosevitskii, M.I., 1968, in: Problems in Molecular Genetics and the Genetics of Microorganisms [in Russian], Moscow, p. 77.
Mosevitskii, M.I., 1969, Vestn. Akad. Nauk SSSR, 8:31.
Mosig, G., 1962, Z. Vererbungsl., 93:280.
Mosig, G., 1967, J. Cell. Comp. Physiol., 70, Suppl. 1:163.
Mosig, G., R. Ehring, and E.D. Duerr, 1968, Cold Spring Harbor Sympos. Quant. Biol., 33:361.
Mosig, G., R. Ehring, W. Schliewen, and S. Bock, 1971, Molec. Gen. Genet., 113:51.
Mosig, G., W. Bowden, and S. Bock, 1972, Nature New Biol., 240:12.
Mosolov, A.N., 1968, Genetika, 6:135.
Mount, D.W., K.B. Low, and S.J. Edmiston, 1972, J. Bacteriol., 112:886.
Mousseau, J., 1966, C. r. Acad. Sci., Sér. D, 262:1254.
Muller, H.J., 1916, Amer. Naturalist, 50:193, 284, 350, 421.
Muller, H.J., 1932, 6th Internat. Congr. Genet., Part I, New York, p. 213.
Muller, H.J., 1964, Mutat. Res., 1:2.
Munakata, N., and Y. Ikeda, 1969, Mut. Res., 7:133.
Mundkur, B.D., 1949, Ann. Missouri Bot. Garden, 36:259.
Munson, R.I., and B.A. Bridges, 1964, Nature, 203:270.
Murray, N.E., 1960, Heredity, 15:207.
Murray, N.E., 1961, Genetics, 46:886.
Murray, N.E., 1963, Genetics, 48:1163.
Murray, N.E., 1968, Genetics, 58:181.
Murray, N.E., 1970, Genet. Res., 15:109.
Nakai, S., and R.K. Mortimer, 1969, Mol. Gen. Genetics, 103:329.
Nance, W.E., 1963, Science, 141:123.

Nash, H.A., and C.A. Robertson, 1971, Virology, 44:446.
Nasim, A., and C. Auerbach, 1967, Mutat. Res., 4:1.
Nebel, B.R., and E.M. Nackett, 1961, Z. Zellforsch., 55:556.
Nelson, O.E., 1962, Genetics, 47:737.
Nester, E.W., M. Schaffer, and I. Lederberg, 1963, Genetics, 48:529.
Nevzgladova, O.V., 1972, Genetika, 8(6):44.
Newcombe, K.D., and S.F.H. Threlkeld, 1972, Genet. Res., 19:115.
Newman, J., and P. Hanawalt, 1968, J. Mol. Biol., 35:639.
Nishioka, H., and C.O. Doudney, 1969, Mut. Res., 8:215..
Nobrega, F.G., F.H. Rola, M. Pasetto-Nobrega, and M. Oishi, 1972, Proc. Nat. Acad. Sci. USA., 65:15.
Nomura, M., and S. Benzer, 1961, J. Mol. Biol., 3:684.
Norkin, L.C., 1970, J. Mol. Biol., 51:633.
Notani, N.K., and J.K. Setlow, 1973, in: Progress in Nucleic Acid Research and Molecular Biology, Academic Press, New York.
Noubo, S., S. Hiuga and J. Yonosuke, 1966, Mutat. Res., 3:93.
Oey, J., and R. Knippers, 1972, J. Mol. Biol., 68:125.
Ogawa, H., 1970, Molec. Gen. Genet., 108:378.
Ohki, M., and J. Tomizawa, 1968, Cold Spring Harbor Sympos. Quant. Biol., 33:651.
Ohshima, S., and M. Sekiguchi, 1972, Biochem. Biophys. Res. Comm., 47:1126.
Oishi, M., 1969, Proc. Nat. Acad. Sci. USA., 64:1292.
Oishi, M., and S.D. Coslov, 1972a, Biochem. Biophys. Res. Comm., 49:1568.
Oishi, M., and S.D. Coslov, 1972b, Proc. Nat. Acad. Sci. USA., 70:84.
Okada, Y., G Streisinger, J.(E.) Owen, J. Newton, A. Tsugita, and M. Inouye, 1972, Nature, 236:338.
Okazaki, R., T. Okazaki, K. Sakabe, S. Kazunori, and G. Sugino, 1968, Proc. Nat. Acad. Sci. USA., 59:598.
Okubo, S., and W.R. Romig, 1966, J. Mol. Biol., 15:440.
Olive, L.S., 1956, Amer. J. Bot., 43:97.
Olive, L.S., 1959, Proc. Nat. Acad. Sci. USA., 45:727.
Oliver, C.P., 1940, Proc. Nat. Acad. Sci. USA., 26:452.
Olivera, B.M., and I.R. Lehman, 1967, Proc. Nat. Acad. Sci. USA., 57:1426.
Opara-Kubinska, Z., H. Kubinski and W. Szybalski, 1964, Proc. Nat. Acad. Sci. USA., 52:923.
Oppenheim, A.B., and M. Riley, 1966, J. Mol. Biol., 20:331.
Oppenheim, A.B., and M. Riley, 1967, J. Mol. Biol., 28:503.
Owen, A.R.G., 1950, Advances Genetics, 3:117.
Painter, I., 1966, J. Mol. Biol., 17:47.
Painter, R.B., and B.R. Young, 1972, Mut. Res., 14:225.
Parchman, L.G., and H. Stern, 1969, Chromosoma, 26:298.
Paribok, V.P., and N.V. Tomilin, 1970, Doka. Akad. Nauk SSSR, 195:489.
Parkinson, J.S., 1971, J. Mol. Biol., 56:385.
Parkinson, J.S., and R.W. Davis, 1971, J. Mol. Biol., 56:425.
Parsons, P. A., 1958, Heredity, 12:77.
Paszewski, A., 1967, Genet. Res., 10:121.
Paszewski, A., 1970, Genet. Res., 15:55.
Paszewski, A., and W. Prazmo, 1969, Genet. Res., 14:33.
Paszewski, A., and S. Surzycki, 1964, Nature, 204:809.
Paszewski, A., S. Surzycki, and M. Mankowska, 1966, Acta Soc. Bot. Polon., 35:181.
Paszewski, A., W. Prazmo, and E. Jaszczuk, 1971, Genet. Res., 18:199.

Pateman, J.A., 1958, Nature, 181:1605.
Pateman, J.A., 1960, Genetics, 45:839.
Patrick, M.H., and C.S. Rupert, 1967, Photochem. Photobiol., 6:1.
Pauling, C., and P. Hanawalt, 1965, Proc. Nat. Acad. Sci. USA., 54:1738.
Pauling, C., S.P. Wilczynski, and J.J. Texera, 1972, Genetics 70 Abstracts, s46.
Peacock, W.J., 1968, in: Replication and Recombination of the Genetic Material, Canberra, p. 24.
Peacock, W.J., 1970, Genetics, 65:593.
Pees, E., 1965, Experientia, 21:514.
Pelling, C., 1968, 12th Internat. Congr. Genet., Part II, Tokyo, p. 74.
Pene, J., and W.R. Romig, 1964, J. Mol. Biol., 9:236.
Perkins, D.D., 1953, J. Cell. Comp. Physiol., 45, Suppl. 2:119.
Perkins, D.D., 1967, J. Cell. Comp. Physiol., 70, Suppl. 1:11.
Perreault, W.J., B.P. Kaufmann, and H. Gay, 1968, Genetics, 60:289.
Peterson, H.M., and J.R. Laughnan, 1963, Proc. Nat. Acad. Sci. USA., 50:126.
Petrell, M.G., and R. Ricci, 1966, Ann. 1st Super. Sanita, 2:379.
Pettijohn, D., and P. Hanawalt, 1964, J. Mol. Biol., 9:395.
Philippe, V., 1966, C. R. Acad. Sci., Sér. D, 273:2010.
Phillips, A.P., 1968, Biochem. Biophys. Res. Commun., 30:393.
Pittard, J., and E.M. Walker, 1967, J. Bacteriol., 96:1656.
Plough, H.H., 1917, Exper. Zool., 24:147.
Plough, H.H., 1924, Amer. Naturalist, 58:654.
Pogosov, V.Z., and S.I. Alikhanyan, 1968, Genetika, 11:168.
Pollard, E., J. Swez, and L. Grady, 1966, Radiat. Res., 28:585.
Pontecorvo, G., 1950, Biochem. Soc. Symp., 4:40.
Pontecorvo, G., 1952, Advances Enzymol., 13:121.
Pontecorvo, G., 1958, Trends in Genetic Analysis, New York.
Pontecorvo, G., and E. Cafer, 1958, Advances Genetics, 9:71.
Pontecorvo, G., and J.A. Roper, 1953, Advances Genetics, 5:218.
Popova, I.A., S.G. Inge-Vechtomov, and E.P. Raipulis, 1968, Genetika, 11:116.
Prakash, V., 1964, Genetica, 35:287.
Prashad, N., and J. Hosoda, 1972, J. Mol. Biol., 70:617.
Prescott, D.M., 1970, Advances Cell Biol., 1:57.
Pritchard, R.H., 1955, Heredity, 9:343.
Pritchard, R.H., 1960a, Symp. Soc. General Microbiol., 10:155.
Pritchard, R.H., 1960b, Genet. Res., 1:1.
Proust, J., and C. Prudhomme, 1968, Mutat. Res., 6:419.
Prozorov, A.A., 1966, Genetika, 8:98.
Prozorov, A.A., and B.I. Barabanshchikov, 1968, Genetika, 5:80.
Prozorov, A.A., R.R. Azizbekyan, B.I. Barabanschikov, and N.N. Belyaeva, 1968, Genetika, 2:97.
Prozorov, A.A., N.A. Kalinin, L.S. Naumov, A.V. Chestuchin, and M.F. Shemyakin, 1972, Genetika, 8(12):142.
Ptashne, M., 1967, Nature, 214:232.
Putrament, A., 1964, Genet. Res., 5:316.
Putrament, A., 1967a, Mol. Gen. Genetics, 100:307.
Putrament, A., 1967b, Mol. Gen. Genetics, 100:321.
Putrament, A., 1971, Genet. Res., 18:85.
Putrament, A., T. Rozbicka, and K. Wojciechowska, 1971, Genet. Res., 17:125.
Radding, C.M., 1970, J. Mol. Biol., 52:491.
Radding, C.M., and D. Carter, 1971, J. Biol. Chem., 246:2513.

Raffel, D., and H.J. Muller, 1940, Genetics, 25:541.
Raipulis, E.P., and Leh Dinh Lyong, 1969, Genetika, 5:129.
Raju, N.B., and B.C. Lu, 1973, Mut. Res., 17:37.
Ravin, A.W., and V.N. Iyer, 1962, Genetics, 47:1369.
Ray, K., L. Bartenstein, and J.W. Drake, 1971, J. Virol., 9:440.
Rayssiguier C., 1972, Lunteren Lectures on Molecular Genetics. S3:4.
Renner, O., 1959, Heredity, 13:283.
Revell, S.H., 1956, in: Problems in Radiobiology [Russian translation], Moscow, p. 388.
Revell, S.H., 1959, Proc. Roy. Soc. B, 150:563.
Rhoades, M.M., 1933, Genetics, 18:535.
Rhoades, M.M., 1947, Genetics, 32:101.
Rhoades, M.M., 1950, J. Heredity, 41:58.
Rhoades, M.M., 1961, in: The Cell, 3, New York–London, p.1.
Rifaat, O.M., 1969, Genetica, 40:89.
Riley, R., 1966, Sci. Progr., 54:193.
Riley, R., and M.D. Bennet, 1971, Nature, 230:182.
Riley, R., and T.E. Miller, 1966, Mutat. Res., 3:355.
Ris, H., 1957, Cold Spring Harbor Sympos. Quant. Biol. 22:16.
Ris, H., and B.L. Chandler, 1963, Cold Spring Harbor Sympos. Quant. Biol., 28:1.
Rizet, G., and C. Engelmann, 1949, Rev. Cytol. et Biol. Végét., 11:201.
Rizet, G., N. Engelmann, C. Lefort, P. Lissouba, and J. Mousseau, 1960, C. r. Acad. Sci., Sér. D, 250:2050.
Rizet, G., and J.L. Rossignol, 1963, Revista de Biologia, 3:261.
Rizet, G., and J.L. Rossignol, 1966, C. r. Acad. Sci., Sér. D, 262:1250.
Robbins, L.G., 1971, Molec. Gen. Genet., 110:144.
Roberts, P.A., 1965, Nature, 205:725.
Roberts, P.A., 1969, Genetics, 63:387.
Rodarte, V., S. Fogel, and R.K. Mortimer, 1968, Genetics, 60:216.
Rodarte-Ramon, V.S., 1972, Rad. Res., 49:148.
Roger, M., 1972, Proc. Nat. Acad. Sci. USA., 69:466.
Roman, H., 1956, Cold Spring Harbor Sympos. Quant. Biol., 21:175.
Roman, H., 1958, Ann. Genet., 1:11.
Roman, H., 1963, in: Methodology in Basic Genetics, San Francisco, p. 209.
Roman, H., 1967, J. Cell. Comp. Physiol., 70, Suppl. 1:116.
Roman, H., and F. Jacob, 1958, Cold Spring Harbor Sympos. Quant. Biol., 23:155.
Roman, H., D.C. Hawthorne, and H.C. Douglas, 1951, Proc. Nat. Acad. Sci. USA., 37:79.
Ronen, A., and Y. Salts, 1971, Virology, 45:496.
Roper, J.A., 1950, Nature, 166:956.
Roper, J.A., 1953, Advances Genetics, 5:208.
Roper, J.A., 1966, in: The Fungi, 2, New York–London, p. 589.
Roper, J.A., and R.H. Pritchard, 1955, Nature, 175:639.
Rosen, H., and W.T. Ebersold, 1972, Genetics, 71:247.
Rosner, J., and N.A. Barricelly, 1967, Virology, 33:425.
Rossen, J.M., and M. Westergaard, 1966, Compt. Rend. Trav. Lab. Carlsberg, Sér. Physiol., 35:233.
Rossignol, J.-L., 1969, Genetics, 63:795.
Rossignol, J.-L., and G. Leblon, 1972, Compt. Rend. Acad. Sci. Paris, Ser. D, 275:3025.
Roth, R., and S. Fogel, 1971, Molec. Gen. Genet., 112:295.
Roth, T.F., 1966, Protoplasma, 61:346.
Roth, T.F., and M. Ito, 1967, J. Cell Biol., 35:247.
Rothfels, K.H., 1952, Genetics, 37:297.

Rottlander, E., K.O. Hermann and R. Hertel, 1967, Mol. Gen. Genetics, 99:34.
Rückert, J., 1892, Anat. Anz., 7:107.
Rupp, W.D., 1967, J. Cell. Comp. Physiol., 70, Suppl. 1:72.
Rupp, W.D., and P. Howard-Flanders, 1968a, J. Mol. Biol., 31:291.
Rupp, W.D., and P. Howard-Flanders, 1968b, Biophys. J., 8, Abstr., A20.
Rupp, W.D., and G. Ihler, 1968, Cold Spring Harbor Sympos. Quant. Biol., 33:647.
Rupp, W.D., C.E. Wilde III, D.L. Reno, and P. Howard-Flanders, 1971, J. Mol. Biol., 61:25.
Rüsh, M.G., and R.C. Warner, 1968, Cold Spring Harbor Sympos. Quant. Biol., 33:459.
Russel, A.E., and E.L. Tatum, 1956, Proc. Nat. Acad. Sci. USA., 42:68.
Russel, K.C., and D. Botstein, 1972, Virology, 49:268.
Russo, V.E.A., 1973, Molec. Gen. Genet.
Rybchin, V.N., 1968, Genetika, 8:87.
Sadowski, P., B. Ginsberg, A. Yudelevich, L. Feiner, and J. Hurwitz, 1968, Cold Spring Harbor Sympos. Quant. Biol., 33:165.
Salamini, F., and C. Lorenzoni, 1970, Mol. Gen. Genetics, 108:225.
Sandler, L., D.L. Lindsey, B. Nicoletti, and G. Trippa, 1968, Genetics, 60:525.
Sarabhai, A.S., A.O. V. Stretton, S. Brenner, and A. Bolle, 1964, Nature, 201:13.
Sasaki, M.S., and A. Norman, 1966, Exper. Cell Res., 44:642.
Savić, D.J., 1972, Genetics, 71:461.
Savić, D.J., and D.T. Kanazir, 1972, Molec. Gen. Genet., 118:45.
Sax, K., 1930, J. Arnold Arboretum, 11:193.
Sax, K., 1931, Science, 74:41.
Scaife, J., 1966, Genet. Res., 8:189.
Scaife, J., and J.D. Gross, 1963, Genet. Res., 4:328.
Schaap, T., 1971, Genetica, 42:219.
Schacht, L.E., 1958, Genetics, 43:665.
Schroeder, A.L., 1968, Genetics, 60:223.
Schwartz, D., 1953, Genetics, 38:251.
Schwartz, D., 1954, Genetics, 39:629.
Searashi, T., and B. Strauss, 1965, Biochem. Biophys. Res. Commun., 20:680.
Sechaud, J., G. Streisinger, J. Emrich, J. Newton, H. Lanford, H. Reinhold, and M.M. Stach, 1965, Proc. Nat. Acad. Sci. USA., 54:1340.
Sedgwick, S.G., and B.A. Bridges, 1972, Molec. Gen. Genet., 119:93.
Serebrovskii, A. S., 1929, Amer. Naturalist, 63:374.
Serebrovskii, A.S., and N.P. Dubinin, 1929, Uspekhi Eksperim. Biol., ser. B, 8(4):64.
Serra, J.A., 1947, Portug. Acta Biol., A2:25.
Serres, F.J. de, 1956, Genetics, 41:668.
Sermonti, G., and A. Carere, 1968, Mol. Gen. Genetics, 103:141.
Setlow, J.K., M.E. Boling, and R.F. Kimball, 1972, J. Mol. Biol., 68:361.
Setlow, R.B., 1964, J. Cell. Comp. Physiol., 64, Suppl. 1:51.
Setlow, R.B., and W. Carrier, 1964, Proc. Nat. Acad. Sci. USA., 51:226.
Setlow, R.B., and J.K. Setlow, 1962, Proc. Nat. Acad. Sci. USA., 48:1250.
Setlow, R.B., P.A. Swenson and W.L. Carrier, 1963, Science, 142:1464.
Setlow, R.B., W.L. Carrier, and J.K. Setlow, 1969a, Biophys. J., 8, Abstr., A57.
Setlow, R.B., J.D. Regan, J. German, and W.L. Carrier, 1969b, Proc. Nat. Acad. Sci. USA., 64:1035.
Shahn, E., 1968, Cold Spring Harbor Sympos. Quant. Biol., 33:205.
Shahn, E., and A. Kozinski, 1966, Virology, 30:455.
Shalitin, C., and F.W. Stahl, 1965, Proc. Nat. Acad. Sci. USA., 54:1340.
Shapot, V.S., 1968, The Nucleases [in Russian], Moscow.

Sharp, P.A., M.T. Hsu, E. Ohtsubo, and N. Davidson, 1972, J. Mol. Biol., 71:471.
Sherman, F., and R. Hershel, 1963, Genetics, 48:255.
Shestakov, S., and S.D. Barbour, 1967, Genetics, 57:283.
Shoemaker, N.B., and W.R. Guild, 1972, Proc. Nat. Acad. Sci. USA., 69:3331.
Sicard, A.M., and H. Ephrussi-Taylor, 1965, Genetics, 52:1207.
Siddiqi, O.H., 1962, Genet. Res., 3:89
Siddiqi, O.H., 1963, Proc. Nat. Acad. Sci. USA., 49:589.
Siddiqi, O.H., and A. Putrament, 1963, Genet. Res., 4:12.
Sidorov, B.N., N.N. Sokolov, and V.S. Andreev, 1965, Genetika, 1:112.
Sidorov, B.N., N.N. Sokolov, and Yu.S. Demin, 1966, Radiobiologiya, 6:84.
Sigal, N., and B. Alberts, 1972, J. Mol. Biol., 72:789.
Sigal, N., H. Delius, T. Kornberg, M.L. Gefter, and B. Alberts, 1972, Proc. Nat. Acad. Sci. USA., 69:3537.
Signer, E.R., 1968, Annual Rev. Microbiol., 22:451.
Signer, E., 1971, in: The Bacteriophage Lambda, Cold Spring Harbor Lab., p. 139.
Signer, E.R., and J.R. Beckwith, 1966, J. Mol. Biol., 22:33.
Signer, E.R., H. Echols, J. Weil, C. Radding, M. Schulman, L. Moore, and K. Manly, 1968, Cold Spring Harbor Sympos. Quant. Biol., 33:711.
Signer, E.R., and V.N. Rybchin, 1967, Genetika, 3:114.
Signer, E.R., and J. Weil, 1968, J. Mol. Biol., 32:261.
Simchen, G., and V. Connolly, 1968, Genetics, 58:319.
Simchen, G., N. Ball, and I. Nachsohn, 1971, Heredity, 26:137.
Simon, E.H., 1961, J. Mol. Biol., 3:101.
Simon, E.H., 1965, Science, 150:760.
Singleton, J.R., 1953, Amer. J. Bot., 40:124.
Skalka, A., 1971, in: The Bacteriophage Lambda, Cold Spring Harbor Lab., p. 535.
Slizynski, B.M., 1964, Genet. Res., 5:80.
Smirnov, G.B., and A.G. Skavronskaya, 1969, Genetika, 1:65.
Smith, D.A., 1961, J. Gen. Microbiol., 24:335.
Smith, D.R., 1971, Austr. J. Biol. Sci., 24:97.
Smith, D.R., and H. Stern, 1972, Genetics 71 Abstracts, s60.
Smith, B.R., 1965, Heredity, 20:257.
Smith, B.R., 1966, Heredity, 21:481.
Smith, H.O., and K.W. Wilcox, 1970, J. Mol. Biol., 51:379.
Smith, K.C., and A.K. Ganesan, 1968, Biophys. J., 8, Abstr., A20.
Smith, P.A., and R.C. King, 1968, Genetics, 60:335.
Smith, P.D., 1972, Genetics 71 Abstracts, s60.
Smith, P.D., V.G. Finnerty, and A. Chovnick, 1970, Nature, 228:444.
Snow, R.J., 1967, J. Bacteriol., 94:571.
Sobell, H.M., 1972, Proc. Nat. Acad. Sci. USA., 69:2483.
Soidla, T.R., 1965, Genetika, 3:127.
Solari, A.J., 1965, Proc. Nat. Acad. Sci. USA., 53:503.
Solari, A.J., 1967, J. Ultrastr. Res., 17:421.
Solari, A.J., and M.J. Moses, 1973, J. Cell Biol., 56:145.
Sparrow, A.H., 1952, Brit. J. Radiology, 25:182.
Sparrow, A.H., and H.I. Evans, 1961, Brookhaven Sympos. Biol., 14:76.
Spatz, H.Ch., and T.A. Trautner, 1970, Mol. Gen. Genetics, 109:84.
Stadler, L.J., 1928, Science, 68:186.
Stadler, D.R., 1956, Genetics, 41:623.
Stadler, D.R., 1959a, Genetics, 44:647.
Stadler, D.R., 1959b, Proc. Nat. Acad. Sci. USA., 45:1625.

Stadler, D.R., 1963, Heredity, 18:233.
Stadler, D.R., and B. Kariya, 1969, Genetics, 63:291.
Stadler, D.R., and A.M. Towe, 1962, Genetics, 47:839.
Stadler, D.R., and A.M. Towe, 1963, Genetics, 48:1323.
Stadler, D.R., and A.M. Towe, 1968, Genetics, 58:327.
Stadler, D.R., and A.M. Towe, 1971, Genetics, 68:401.
Stadler, D.R., A.M. Towe, and J.-L. Rossignol, 1970, Genetics, 66:429.
Stahl, F.W., 1961, J. Chim. Phys. et Phys.-Chim. Biol., 58:1072.
Stahl, F.W., 1965, The Mechanics of Heredity, Englewood Cliffs, N.J. (Russian translation, Moscow, 1966).
Stahl, F.W., 1967, J. Cell. Comp. Physiol., 70, Suppl. 1:1.
Stahl, F.W., 1969, Genetics, 61, 1(Suppl):1.
Stahl, F.W., and M.M. Stahl, 1971, in: The Bacteriophage Lambda, Cold Spring Harbor Lab., p. 443.
Stahl, F.W., and C.N. Steinberg, 1964, Genetics, 50:531.
Stahl, F.W., R.S. Edgar and J. Steinberg, 1964, Genetics, 50:539.
Stahl, F.W., H. Modersohn, B.E. Terraghi, and J.M. Crassemann, 1965, Proc. Nat. Acad. Sci. USA., 54:1342.
Stahl, F.W., K.D. McMilin, M.M. Stahl, R.E. Malone, Y. Nozu, and V.E.A. Russo, 1972a, J. Mol. Biol., 68:57.
Stahl, F.W., K.D. McMilin, M.M. Stahl, and Y. Nozu, 1972b, Proc. Nat. Acad. Sci. USA., 69:3598.
Stahl, F.W., S. Chung, J. Grasemann, D. Faulds, J. Haemer, S. Lam, R.E. Malone, K.D. McMilin, Y. Nozu, J. Siegal, J. Strathern, and M.M. Stahl, 1973, Personal communication.
Stahl, M.M., and F.W. Stahl, 1971, in: The Bacteriophage Lambda, Cold Spring Harbor Lab., p. 431.
Steinberg, C.M., and R.S. Edgar, 1962, Genetics, 47:187.
Steinberg, C.M., and F.W. Stahl, 1967, J. Cell. Comp. Physiol., 70, Suppl. 1:13.
Steinhart, W.L., and R.M. Herriott, 1968, J. Bacteriol., 96:1718.
Stent, G., 1963, Molecular Biology of Bacterial Viruses, San Francisco and London. (Russian translation, Moscow, 1965).
Stern, C., 1931, Biol. Zbl., 51:547.
Stern, C., 1933, Faktorenkoppelung und Faktorenaustausch, Berlin.
Stern, C., 1936, Genetics, 21:625.
Stevens, W.L., 1936, J. Genetics, 32:51.
Storm, P.K., W.P.M. Hoekstra, P.G. de Haan, and C. Verhoef, 1971, Mut. Res., 13:9.
Strauss, B., M. Coyle, and M. Robbins, 1968, Cold Spring Harbor Sympos. Quant. Biol., 33:277.
Strauss, N., 1970, Genetics, 66:583.
Streisinger, G., R.S., Edgar, and G.H. Denhardt, 1964, Proc. Nat. Acad. Sci. USA., 51:775.
Strickland, W.N., 1958a, Proc. Roy. Soc. B, 148:533.
Strickland, W.N., 1958b, Proc. Roy. Soc. B, 149:82.
Strickland, W.N., 1961, Genetics, 46:1125.
Stubblefield, E., and W. Wray, 1971, Chromosoma, 32:262.
Sturtevant, A.H., 1913, J. Exper. Zool., 14:43.
Sturtevant, A.H., 1925, Genetics, 10:117.
Sturtevant, A.H., 1951, Proc. Nat. Acad. Sci. USA., 37:405.
Sturtevant, A.H., 1965, Amer. Scientist, 53:303.
Stuy, J.H., J.E. Hoffmann, and L.H. Duket, 1972, Genetics, 71:507.
Sueoka, N., K.S. Chiang, and J.R. Kates, 1967, J. Mol. Biol., 25:47.

Sugimoto, K., T. Okazaki, and R. Okazaki, 1968, Proc. Nat. Acad. Sci. USA., 60:1356.
Sukhodolets, V.V., 1965, Genetika, 12:26.
Sukhodolets, V.V., and V.P. Galeis, 1968, Genetika, 12:26.
Sukhodolets, V.V., T.S. Il'ina, and S.I. Alikhanyan, 1965, Genetika, 1:78.
Sunderland, N., and J. McLeish, 1961, Exper. Cell Res., 24:541.
Sutton, W.S., 1902, Biol. Bull. Mar. Biol. Lab., Woods Hole, 4:24.
Sutton, W.S., 1903, Biol. Bull. Mar. Biol. Lab., Woods Hole, 4:231.
Suyama, Y., K.D., Munkers, and V.W. Woodward, 1959, Genetics, 30:293.
Suzuki, D.T., 1963a, Genetics, 48:1605.
Suzuki, D.T., 1963b, Canad, J. Genet. Cytol., 5:482.
Suzuki, D.T., 1965, Genetics, 51:635.
Sved, J.A., 1966, Genetics, 53:747.
Swietlinska, Z., and H.J. Evans, 1970, Mutat. Res., 10:185.
Swift, H., 1950, Proc. Nat. Acad. Sci. USA., 36:643.
Symonds, N., 1962, Virology, 18:334.
Symonds, N., H. Heindl, and P. White, 1973, Molec. Gen. Genet., 120:253.
Szybalski, W., 1964, Abhandl. Dtsch. Akad. Wiss. Berlin, Kl. Med. Wiss., 4:1
Szybalski, W., and Z. Lorkiewicz, 1962, Abhandl. Dtsch. Akad. Wiss. Berlin, Kl. Med. Wiss., 1:63.
Taketo, A., S. Yasuda, and M. Sekiguchi, 1972, J. Mol. Biol., 70:1.
Taylor, J.H., 1953, Exper. Cell Res., 4:164.
Taylor, J.H., 1957, Amer. Naturalist, 91:209.
Taylor, J.H., 1958, 10th Internat. Congr. Genet., Part I, Montreal, p. 63.
Taylor, J.H., 1960, Cancer Res., 14:547.
Taylor, J.H., 1963, in: Molecular Genetics, Vol. 1, New York and London. (Russian translation, Moscow, 1964, p. 78).
Taylor, J.H., 1965, J. Cell Biol., 25:57.
Taylor, J.H., 1967, in: Molecular Genetics, Vol. 2, New York and London, p. 95.
Taylor, J.H., 1973, Proc. Nat. Acad. Sci. USA., 70:1083.
Taylor, J.H., and S.H. Taylor, 1953, J. Heredity, 44:128.
Taylor, J.H., W.F. Haut, and J. Tung, 1962, Proc. Nat. Acad. Sci. USA., 48:190.
Templin, A., S.R. Kushner, and A. J. Clark, 1972, Genetics, 72:205.
Tessman, I., 1965, Genetics, 51:63.
Thoday, J.U., 1954, New Phytologist, 53:511.
Thomas, C.A., 1966, Progr. Nucl. Acid. Res., 5:315.
Thomas, C.A., 1967, J. Cell. Comp. Physiol., 70, Suppl. 1:13.
Thomas, C.A. Jr., 1971, Ann. Rev. Genetics, 5:237.
Thomas, P.L., and D.G. Catcheside, 1969, Canad. J. Genetics and Cytol., 11:558.
Thuriaux, P., M. Minet, A.M.A. fon Berge, and F. K. Zimmermann, 1971, Molec. Gen. Genet., 112:60.
Tomilin, N.V., 1970, in: Proceedings of a Scientific Conference of the Institute of Cytology, Academy of Sciences of the USSR [in Russian], Leningrad, p. 84.
Tomizawa, J., 1960, Proc. Nat. Acad. Sci. USA., 46:91.
Tomizawa, J., 1967, J. Cell. Comp. Physiol., 70, Suppl. 1:201.
Tomizawa, J., and N. Anraku, 1964, J. Mol. Biol., 8:508.
Tomizawa, J., and H. Ogawa, 1972, Nature New Biol., 239:14.
Tomizawa, J., N. Anraku, and Y. Iwama, 1966, J. Mol. Biol., 21:247.
Touré, B., 1972a, Lunteren Lectures on Molecular Genetics, S1:3.
Touré, B., 1972b, Molec. Gen. Genet., 117:267.
Touré, B., and M. Picard, 1972, Genet. Res., 19:313.
Town, C.D., K.C. Smith, and H.S. Kaplan, 1971, Science, 172:851.

Trautner, T.A., 1958, Z. Vererbungsl., 89:264.
Trosko, J.E., and K. Wilder, 1973, Genetics, 73:297.
Trow, A.H., 1913, J. Genetics, 2:313.
Tuveson, R.W., 1969, Amer. Naturalist, 103:23.
Uhl, C.H., 1965, Genetics, 51:191.
Unger, R.C., and A.J. Clark, 1972, J. Mol. Biol., 70:539.
Unger, R.C., H. Echols, and A.J. Clark, 1972, J. Mol. Biol., 70:531.
Unrau, P., and R. Holliday, 1972, Genet. Res., 19:145.
Val'dshtein, E.A., and V.D. Zhestyanikov, 1966, in: Protection and Regeneration in Radiation Injuries [in Russian], Moscow, p. 5.
Valentin, J., 1972, Hereditas, 72:243.
Vapnek, D., and W.D. Rupp, 1970, J. Mol. Biol., 53:287.
Venema, G., R.H. Pritchard, and T. Venema-Schröder, 1965, J. Bacteriol., 90:343.
Verhoef, G., and P.G. de Haan, 1966, Mutat. Res., 5:101.
Vielmetter, W., F. Bonhoeffer, and A. Schutte, 1968, J. Mol. Biol., 37:81.
Vinetskii, Yu.P., and N.M. Aleksandrova, 1972, Genetika, 8(9):140.
Visconti, N., and M. Delbrück, 1953, Genetics, 38:5.
Vol'kenshtein, M.V., 1965, Molecules and Life [in Russian], Leningrad.
Voll, M.J., and S.H. Goodgal, 1961, Proc. Nat. Acad. Sci. USA., 47:505.
Vovis, G.F., 1973, J. Bacteriol., 113:718.
Vovis, G.F., and G. Buttin, 1970, Biochim. Biophys. Acta, 224:42.
Wackernagel, W., 1973, Biochem. Biophys. Res. Comm., 51:306.
Walmsley, R.H., 1969, Biophys., J., 9:421.
Walters, J. L., 1956, Amer. J. Bot., 43:342.
Walters, M.S., 1968, Genetics, 60, 1, Part 2:235.
Watson, J.D., 1972, Nature New Biol., 239:197.
Watson, J.D., and F.H.C. Crick, 1953a, Cold Spring Harbor Sympos. Quant. Biol., 18:123.
Watson, J.D., and F.H.C. Crick, 1953b, Nature, 171, 964.
Watson, G.S., W.K. Smith, and C.A. Thomas, 1966, Progress in Nucleic Acid Research, 5:315.
Weil, J., 1969, J. Mol. Biol., 43:351.
Weil, J., and E.R. Signer, 1968, J. Mol. Biol., 34:273.
Weinstein, A., 1936, Genetics, 21:155.
Weinstein, A., 1958, Cold Spring Harbor Sympos. Quant. Biol., 23:177.
Weintraub, S.B., and F.R. Franklin, 1972, J. Mol. Biol., 70:589.
Weismann, A., 1885, Die Kontinuität des Keimplasmas, als Grundlage einer Theorie der Vererbung, Jena.
Weismann, A., 1887, Ueber die Zahl der Richtungskorper und ihre Bedeutung fur die Verebung, Jena.
Weiss, B., and C. C. Richardson, 1967, Proc. Nat. Acad. Sci. USA., 57:1027.
Welshons, W.I., and E.S. von Halle, 1962, Genetics, 47:743.
Wenrich, D., 1916, Bull. Museum Compar. Zool. Harv., 60:57.
Westergaard, M., 1964, Compt. Rend. Trav. Lab. Carlsberg, Sér. Physiol., 34:359.
Westergaard, M., and D. von Wettstein, 1970, C.R. Trav. Lab. Carlsberg, 37:239.
Westergaard, M., and D. von Wettstein, 1972, Ann. Rev. Genetics, 71.
Westmoreland, B.C., W. Szybalski, and H. Ris, 1969, Science, 163:1343.
Wettstein, D. von, 1971, Proc. Nat. Acad. Sci. USA., 68:851.
Whitehouse, H.L.K., 1963, Nature, 199: 1034.
Whitehouse, H.L.K., 1965, Towards an Understanding of the Mechanism of Heredity, London.
Whitehouse, H.L.K., 1966, Nature, 211:708.
Whitehouse, H.L.K., 1967a, Nature, 215:1352.

Whitehouse, H.L.K., 1967b, J. Cell Sci., 2:9.
Whitehouse, H.L.K., 1969, Towards an Understanding of the Mechanism of Heredity, 2nd ed., E. Arnold, London, p. 328.
Whitehouse, H.L.K., 1970, Biol. Rev., 45:265.
Whitehouse, H.L.K., 1972, Brookhaven Symposia in Biology, 23:293.
Whitehouse, H.L.K., and P.J. Hastings, 1965, Genet. Res., 6:27.
Whittinghill, M., 1955, J. Cell. Comp. Physiol., 45, Suppl. 2:189.
Wiemann, J., 1965, Z. Vererbungsl., 97:81.
Wildenberg, J., 1970, Genetics, 66:291.
Wilkins, B.M., 1968, J. Gen. Microbiol., 51:107.
Wilkins, B.M., 1969, J. Bacteriol., 98:599.
Wilkins, B.M., and P. Howard-Flanders, 1968, Genetics, 60:243.
Wilkins, B.M., S.E. Hollom., and W.D. Rupp, 1971, J. Bacteriol., 107:505.
Willetts, N.S., 1972, Lunteren Lectures on Molecular Genetics, S3:4.
Willetts, N.S., and A.I. Clark, 1969, J. Bacteriol., 100:231.
Wimber, D.E., and W. Prensky, 1963, Genetics, 48:1731.
Winge, Ö., 1955, Heredity, 9:373.
Winkler, H., 1930, Die Konversion der Gene, Jena.
Winkler, H., 1932, Biol. Zbl., 52:163.
Wolfe, S.L., and J.N. Grim., 1967, J. Ultrastr. Res., 19:382.
Wolff, S., 1969, Internat. Rev. Cytol., 25:279.
Wood, T.H., 1967, Science, 157:319.
Wood, T.H., 1969, Biophys. J., 9, Abstr., A253.
Worgel, A., and E. Burgi, 1972, J. Mol. Biol., 71:127.
Worthy, T.E., and J.L. Epler, 1972, Genetics 71 Abstracts, s69.
Wright (Wallace), M.E., 1947, Heredity, 1:349.
Wright, M., G. Buttin, and J. Hurwitz, 1971, J. Biol. Chem., 246:6543.
Wu, T.T., 1967, J. Theoret. Biol., 17:40.
Wulff, D.L., and C.S. Ruppert, 1962, Biochem. Biophys. Res. Commun., 7:237.
Yamamoto, N., 1967, Biochem. Biophys. Res. Commun., 27:263.
Yanofsky, C., 1960, Bacteriol. Revs. 24:221.
Yanofsky, C., B.C. Carlton, J.R. Guerst, D.R. Helinski, and V. Henning, 1964, Proc. Nat. Acad. Sci. USA., 51:266.
Yasuda, S., and M. Sekiguchi, 1970a, J. Mol. Biol., 47:243.
Yasuda, S., and M. Sekiguchi, 1970b, Proc. Nat. Acad. Sci. USA., 67:1839.
Yost, H.T., and R.N. Benneyan, 1957, Genetics, 42:147.
Youngs, D.A., and I.A. Bernstein, 1973, J. Bact., 113:901.
Youngs, D.A., and K.C. Smith, 1973, J. Bacteriol., 114:121.
Zadražil, S., and V. Fučik, 1971, Biochem. Biophys. Res. Comm., 42:676.
Zakharov, I.A., and S.G. Inge-Vechtomov, 1961, Research in Genetics, Collection 1 [in Russian], Leningrad University Press, p. 25.
Zakharov, I.A., and K.V. Kvitko, 1967, The Genetics of Microorganisms [in Russian], Leningrad.
Zakharov, I.A., and T.N. Kozhina, 1968, Genetika, 9:59.
Zavil'gel'skii, G.B., 1965, in: Molecular Biophysics [in Russian], Moscow, p. 137.
Zen, S., 1964, Cytologia, 26:57.
Zhdanov, V.M., and V.M. Stakhanova, 1965, Genetika, 6:47.
Zhestyanikov, V.D., 1969, Regeneration and Radioresistance of the Cell [in Russian], Leningrad.
Zickler, H., 1934, Planta, 22:573.
Zimmermann, F.K., 1968, Mol. Gen. Genetics, 101:171.

Zimmerman, F.K., 1971, Mut. Res., 11:327.
Zimmermann, F.K., and U. von Laer, 1967, Mutat. Res., 4:377.
Zimmermann, F.K., and R. Schwaier, 1967, Mol. Gen. Genetics, 100:63.
Zimmerman, S.B., J.W. Little, C.K. Oshinsky, and M. Gellert, 1967, Proc. Nat. Acad. Sci. USA., 57:1841.
Zipser, D., 1967, Science, 157:1176.
Zlotnikov, K.M., and M.I. Khmelnitskii, 1973, Doklady Akad. Nauk SSSR, 209:717.